The Diffusion of Innovations

This book is part of the Peter Lang Media and Communication list.
Every volume is peer reviewed and meets
the highest quality standards for content and production.

PETER LANG
New York • Washington, D.C./Baltimore • Bern
Frankfurt • Berlin • Brussels • Vienna • Oxford

The Diffusion of Innovations

A Communication Science Perspective

Edited by Arun Vishwanath and George A. Barnett

PETER LANG
New York • Washington, D.C./Baltimore • Bern
Frankfurt • Berlin • Brussels • Vienna • Oxford

Library of Congress Cataloging-in-Publication Data

The diffusion of innovations: a communication science perspective /
edited by Arun Vishwanath, George A. Barnett.
p. cm.
Includes bibliographical references and index.
1. Diffusion of innovations. 2. Communication. 3. Technological innovations.
I. Vishwanath, Arun. II. Barnett, George A.
HM101.D536 303.48'3—dc22 2010046074
ISBN 978-1-4331-1083-2 (hardcover)

Bibliographic information published by **Die Deutsche Nationalbibliothek**.
Die Deutsche Nationalbibliothek lists this publication in the "Deutsche
Nationalbibliografie"; detailed bibliographic data is available
on the Internet at http://dnb.d-nb.de/.

29 Broadway, 18th floor, New York, NY 10006
www.peterlang.com

To Vera, Zachary and Sarah, with love

CONTENTS

Introduction

ARUN VISHWANATH & GEORGE A. BARNETT

Everett Rogers's (2003) diffusion of innovation theory details the process by which a new innovation—a product, practice, or idea—diffuses through a social system. It is a middle-range theory (Merton, 1968), one that provides empirically derived and testable propositions about the generic diffusion process, its constituents, antecedents, and consequences, all of which can be easily adapted to advance the study of the diffusion of any innovation.

In diffusion theory, new innovations are thought to create uncertainty among individuals because the outcomes of adopting the innovation are unknown to its potential adopters. This uncertainty motivates individuals to seek subjective information about the innovation from near peers and others in their social network (Valente, 1994). A socially constructed meaning of the innovation emerges as individuals exchange information and converge on a shared understanding of the innovation (Barnett & Siegel, 1988; Fulk, 1993; Rogers & Kincaid, 1981). Hence, the process of information exchanged through communication is central to the diffusion of innovations.

In the last few years, diffusion scholarship has focused less on the process of diffusion and the processes that influence adoption and more on factors outside of communication, such as the attributes of the innovation, that influence adoption. Another important change is in the topic of study. Early research on the dissemination of news, ideas, information, culture, networks, and health behavior is no longer the focus of diffusion research. Rather, the current focus is generally limited to the diffusion of new technologies. Another change has been in how communication scholars approach the study of adoption. Today's diffusion scholars draw heavily from the Management Information Science (MIS) perspective, utilizing MIS models such as the Technology Acceptance Model (Davis, Bagozzi & Warshaw, 1989), relying on a-priori measures and scales and focusing on the individual's likelihood to adopt the innovation. Moreover, diffusion research continues to focus on the likelihood of adoption of often a single innovation instead of technology clusters or competing innovations (Vishwanath & Chen, 2006). Finally, most research ignores the possibility of the rejection of innovations and discounts the dis-adoption process. In short, diffusion scholarship continues to suffer from a pro-innovation bias, an issue Rogers belabored for over two decades.

The pro-innovation bias is the implication that an innovation should be diffused and adopted by all members of society, that it should be diffused more rapidly, and that it should be neither re-invented nor rejected or its use discontinued (Rogers, 2003). Diffusion research has often been funded by government agencies or private corporations with a vested interest in successfully marketing their innovation in as short a time frame as possible either to maximize profits or to achieve policy objectives. Further, diffusion scholars recognize that innovations are reinvented, changed, and modified by users during the process of adoption or implementation. Rice and Rogers (1980) provide many examples of such reinventions of innovations. These changes and modifications take place because the meanings of innovations are negotiated by adopters through their interactions with the interpersonal and mass-mediated messages about the innovation. These messages result in a collective frame that emerges and influences the subsequent adopters of the innovation as well as their ultimate experience with the innovation (Vishwanath, 2009).

In his preface to the last edition of the *Diffusion of Innovations*, Rogers (2003) called for less of the same in diffusion research and a move away from simple models of explanation. The challenge for diffusion scholars, he noted, was to move beyond the proven methods and models of the past, recognize the limitation of prior approaches, and broaden the conceptions of the diffusion of innovations.

This book is an answer to that call. The book brings together some of the most noted scholars from the field of communication, many of whom directly or indirectly worked with Everett Rogers, and who have over the years contributed to our broader understanding of the diffusion process.

In the first chapter Arun Vishwanath and Hao Chen critically examine the extant use of linear models to predict the intent to adopt a singular innovation using just a handful of predictor variables. In this chapter, the authors present a relational model of adoption that compares an adopter's relative choice among multiple innovations. The model is built on an associational framework of cognition, a concept that has been around since the time of Aristotle, where concepts are seen as related to each other through associative chains. The relational model is empirically tested using a metric multidimensional scaling system that is particularly well suited for such measurement. When compared with the traditional approaches, the relational model of choice provides the probabilities associated with the choice of an innovation from a basket of other innovations, along with the attributes that drive this choice. In doing so the model provides a much more comprehensive and accurate assessment of the innovation-decision process.

In the second chapter, Oscar Peters compares three generally accepted models in the field of communication science: the expectancy-value perspective on uses and gratifications, the social cognitive perspective on communication technology adoption, and the unified theory of acceptance and use of technology. The convergence between the three perspectives is examined in terms of how useful they are in their ability to explain adoption. The chapter argues that the choice of a model to understand a particular technology adoption behavior should be determined foremost by the stage of development and diffusion the particular technology is in.

In the third chapter, Veronika Karnowski, Thilo von Pape, and Werner Wirth examine the shortcomings in diffusion research, particularly its views of adopters and what they do with innovations. Their chapter responds to this shortcoming on several levels: by describing how mainstream diffusion theory, from its earliest days, has considered adoption from a binary perspective; by explaining the methodological and theoretical reasons for this approach; by showing how the phenomenon of re-invention has always challenged this view, even more so today with the increasing complexity of innovations; and finally by outlining approaches that may lead to a new and broader perspective on what users do with innovations in the course of the diffusion process.

In the fourth chapter, Lidwien van de Wijngaert and Harry Bouwman argue for the need to take a user perspective in diffusion research and study the interaction between the emergence of needs and the use of services along with the influence of time on this process. The authors provide a process model of technology adoption and demonstrate an alternative research approach to test their model. The research approach draws from factorial surveys, policy capturing, and vignette studies on the idea of presenting individuals with hypothetical technology adoption and use situations. These scenarios, vignettes, or cases are developed by combining characteristics of different possible situations that vary in contextual, situational, and technical factors. Tracking user perceptions and potential reactions to different vignettes provides a richer, more detailed understanding of the user acceptance process.

In the fifth chapter, George Barnett challenges scholars who study the diffusion of innovation over time to move beyond the S-shaped curve. This traditional conceptualization of the diffusion process is an oversimplification and is based on a number of assumptions about the adopting population, the channels through which information is disseminated, and the characteristics of the innovation. The chapter reviews various mathematical models that have been used to describe the diffusion process and discusses the assumptions on which they are based. It then suggests alternative models that make different assumptions and which may provide better descriptions of the diffusion process such as what channels are used to diffuse the innovation and whether or not the use of the innovation is discontinued.

In the sixth chapter, James Danowski, Julia Gluesing, and Ken Riopelle also question the S-shaped diffusion curve that is a consequence of interpersonal discussions among adopters during the spread of an innovation. The chapter argues that with new media come a different set of assumptions about the key variables that grounded the old S-shaped curve. A new form of curve emerges, the convex curve, one shaped similarly to the trajectory of a rocket launched into low-earth orbit. This chapter presents the rationale underlying the revolution in diffusion theory wrought by new media technologies. The more slowly growing S-shaped curves still explain certain types of innovation diffusion, but they are increasingly relegated to a more circumscribed position in the theoretical space.

In the seventh chapter, Frank Tutzauer, Kyounghee Hazel Kwon, and Benjamine Elbirt demonstrate the utility of exploring social network effects on the diffusion of innovations using agent-based modeling (ABM). Because of the difficulty in collecting and tracking influence information from all pairs of individuals within large systems over time, social network research on diffusion that uses empirical data is often limited to cross-sectional studies of small-sized networks. ABM is an alternative to the empirical approach and allows researchers to simulate agents' behaviors and observe how these micro behaviors influence the system as a whole. Using ABM, the authors compare the influence of opinion leadership, structural equivalence, and time since latest adoption on the diffusion of two competing ideas within three different types of real-life social networks, a Facebook friendship network, an advice network, and a network of jazz bands.

In the eighth chapter, Carolyn Lin examines the phenomenon of media substitution and provides a theoretical foundation for a more accurate forecasting of the phenomenon. The chapter argues that the media substitution phenomenon is neither simple nor straightforward, as the phenomenon contains a content, technical, and social dimension. Fundamentally, the popularity of cotemporary entertainment and communication media technologies is now determined by how well the functions and applications of the technology can accommodate a user-centered design, where the user interfaces with the technology and the content it delivers needs to reflect strong user satisfaction in terms of technical usability and content usefulness. The technology that provides the best match between desirable media content and an efficient delivery modality in an economic manner will typically survive.

In the ninth chapter, Arvind Singhal presents an alternative conceptualization of diffusing innovations termed as the Positive Deviance (PD) approach. The PD approach expands the theoretical space of solutions for diffusing innovations by working with a different set of principles, questions, and mindsets, believing that often the wisdom to solve intractable social problems lies within the community. Diffusion in the PD approach is an inside-out process in contrast to the classical dominant framework of outside-in diffusion. The chapter describes the Positive Deviance approach, including its key tenets and principles, by analyzing its historical origins in Vietnam to combat endemic malnutrition. Through the experience of this pioneering real-life application of PD in Vietnam, and drawing upon dozens of other applications that have followed, the author argues for an alternative conceptualization of diffusion of innovations—one that turns upside-down our cherished conceptualizations of expert and outside change agents, the notion of filling knowledge-attitude-practice (KAP) gaps, and the traditional role of opinion leaders.

In the tenth chapter, James Dearing and Gary Meyer make a case for a user-centric understanding of personal agency or adopter activity during the diffusion process within the domain of translational research. They argue that adopters of innovations are by no means passive receivers of innovations. Rather, they are extremely active in testing, manipulating, and doing what it takes to negotiate and create an innovation that precisely addresses the requirements of a local problem. If the goal of successful diffusion is to heighten the utility of a given innovation and hence transform it from something that one will not adopt or use into something that suits a felt need well enough to cause both adoption and implementation, then understanding adopter activity is vital. The chapter therefore argues for a decentralized view of the diffusion process—one that accounts for the heterogeneity of adopter activity, influence, and choices throughout the process of diffusion.

In many ways, the chapters in this book expose what is missing in diffusion research and open new theoretical and methodological frontiers for research in diffusion. The book is a call to future diffusion scholars to rethink the fundamental assumptions of the theory and reconsider the limitations of present approaches. Future diffusion scholars are hence an audience for this book, as are marketers of ideas and products, communication and management consultants, policy makers, and individuals and organizations working on changing the status quo within social systems. The book is also an indispensible resource for anyone wishing to study the current trends in the study of the diffusion of innovations since the publication the fifth edition of Rogers's (2003) diffusion book. The book is therefore recommended as a supplementary text for upper-level undergraduate and graduate courses in communication, business and management, sociology, information science, and sociology.

In summary, each chapter advances our understanding of the diffusion process. Using approaches ranging from multidimensional scaling to agent-based modeling, each chapter critically examines the extant theoretical and methodological approaches and perspectives in diffusion research and presents compelling new ways to understand the process of diffusion. In doing so, each chapter expands the scope of diffusion theory and lays the groundwork for the next generation of scholarship in the field of communication.

References

Barnett, G. A. & Siegel, G. (1988). The diffusion of computer assisted legal research systems. *Journal of the American Society for Information Science*, 39, 224–234.

Davis, F. D., Bagozzi, R. P. & Warshaw, P. R. (1989). User acceptance of computer technology: A comparison of two theoretical models. *Management Science*, 35, 982–1003.

Fulk. J. (1993). Social construct of communication technology. *Academy of Management Journal*, 36, 821–950.

Merton, R. K. (1968). *Social theory and social structure*. New York: Free Press.

Rice, R. E., & Rogers, E. M. (1980). Re-Invention in the innovation process. *Knowledge*, *1*, 499–514.

Rogers, E. M. (2003). *Diffusion of innovations* (5th ed.). New York: Free Press.

Rogers, E. M., & Kincaid, D. L. (1981). *Communication networks: Toward a new paradigm for research.* New York: Free Press.

Valente, T. W. (1994). *Network models of the diffusion of innovations*. Cresskill, NJ: Hampton.

Vishwanath, A. (2009). From beliefs to intentions: The impact of framing on the adoption of innovations. *Communication Monographs*, *76*, 177–206.

Vishwanath, A., & Chen, H. (2006). Technology clusters: Using multidimensional scaling to evaluate and structure technology clusters. *Journal of the American Society for Information Science & Technology*, *57*, 1451–1460.

· 1 ·

Towards a Comprehensive Understanding of the Innovation-Decision Process

A Relational Model of Adopter Choice

ARUN VISHWANATH & HAO CHEN

Introduction

In almost all decision-making scenarios, individuals choose among multiple alternatives and utilize a personal valuation scheme to cognitively process and reduce the set of alternatives to one final choice. This basic facet of human decision-making is ignored by almost all models of innovation adoption. Most current approaches to modeling adoption focus on the individuals' decision to adopt a singular innovation that is preselected for them by the researcher prior to model testing. Moreover, these approaches assume that the valuations used by the adopters remain stable across time and space. In this chapter, we present a relational model of adoption, one that accounts for the adopters' relative choice among different innovations and the evaluative schema used to arrive at a final choice. The model is built on an associational framework of cognition, a concept that has been around since the time of Aristotle, where concepts are seen as related to each other through associative chains. The relational model provides the probabilities associated with the choice of an innovation from a basket of other innovations, along with the perceived attributes that drive this choice. In doing so the model provides a much more comprehensive and accurate assessment of the innovation-decision process.

Diffusion theory presents a demand side explanation of how and why an innovation diffuses through a social system (Rogers, 2003). The theory presents a broad framework that encompasses all the factors that could influence individual adoption and thereby the rate at which an innovation diffuses through a social system. The framework includes as predictors proximate factors such as change agent efforts and the adopters' volitional control over their adoption decision along with their perceptions of the attributes of the innovation. It also includes ultimate factors such as cultural norms and the types of communication channels that exist in the system.

Among the proximate factors, the perceived attributes of the innovation account for the majority of variance in the rate of adoption of an innovation (Rogers, 2003). Five attributes of innovations are considered key to a potential adopters' decision to adopt: the relative advantage of the innovation over preceding innovations; the compatibility of the innovation with other existing technologies; the complexity in learning to use the innovation; the perceived visibility or ability to vicariously observe its consumption; and the ability to test the innovation on a limited basis.

Given the relatively high influence of perceived innovation attributes on adoption, contemporary research has focused on just these factors and their impact on the individuals' decision to adopt an innovation or behavioral intent. To their credit, this research has developed precise measures (based on alpha reliability that exceeds 0.90 in many cases) and designed strong measurement models that consistently explain over 70% of the variance in the individual's intent to adopt an innovation, using just a handful of independent measures. The amount and quality of work has been so precise that many scholars have questioned the need to further study diffusion. Thus, it appears that diffusion research has reached a saturation point.

The goal of this chapter is to evaluate whether this is truly the case. Have we reached a point where we can successfully predict whether an individual would adopt an innovation? Can we comprehensively explain how an individual arrives at an adoption decision, or are we just able to explain a subset of a larger process? In this chapter, we argue for the latter. We contend that innovations exist in a spectrum of choice and adopters usually compare the relative value of competing innovations while deciding on the innovation they intend to adopt. We further argue that this comparative choice process varies depending on the adopters' personality and other proximate factors. These important facets of the adoption decision process are ignored by the current models of adoption, which instead focus on a singular innovation and the absolute perceptions of adopters regarding that innovation. We believe that this is because of the limitations of present approaches to empirical measurement that have in turn resulted in an incomplete and inaccurate representation of the adoption decision process. Hence, neither the process of adoption nor the individual decision to adopt an innovation is well understood.

In this chapter we argue for a focus on the adopters' choice among innovations instead of their absolute behavioral intent and present a comprehensive approach to modeling this choice process. We begin the chapter with an examination of the present approaches to modeling the decision to adopt an innovation followed by the limitations of these approaches. Next we present a relational model of adopter choice that we believe better represents the innovation choice process, followed by a methodology that is best suited to test this model. Finally, we present an empirical examination of our relational model along with conclusions and implications for future diffusion research.

Models of Adoption

The extant models that explain the individuals' decision to adopt an innovation can be broadly classified based on their measurement and modeling approaches. One approach is to derive the proximate measures that predict adoptive intent on a study specific basis. This can be thought of as a traditional modeling approach and is rarely used in current diffusion research. This approach drew from the social-psychological tradition of behavioral measurement, espoused by models such as Fishbein and Ajzen's (1975) Theory of Reasoned Action (TRA), and was advocated by Rogers (1964). Researchers in this tradition would derive their measures using focus groups or other projective techniques and utilize factor analysis and OLS regression techniques to refine the measures and accord relative value weights to the independent measures (e.g., Leung & Wei, 1999).

This approach, though exploratory, had three inherent advantages. First, the measures that were derived had high content validity because they were drawn directly from the study population. Second, this approach could be easily extended to study the diffusion of any technology or innovation. Third, researchers would often include and at other times uncover additional constructs during the exploratory process and therefore the approach provided a richer understanding of the adoption decision process.

The second, more contemporary approach to modeling adoption is to utilize a-priori scales and measures. This approach is confirmatory and employs structural equation modeling and other techniques aimed at validating the fit between empirically derived data and a predefined research model. A root of this approach can be found in the Technology Acceptance Model (TAM) (Davis, Bagozzi & Warshaw, 1989) from the management information sciences (MIS). The model is presented in Figure 1. TAM was built to explain the acceptance of computing technology by end users within organizations. The model draws from the TRA framework and distinguishes between the perceived beliefs about software-based innovations and the end users' behavioral intent. According to TAM, the end users' intent to accept an innovation is best predicted by just two instrumental attributes of innovations: their perceived ease of use and perceived usefulness. Perceived ease of use refers to the degree of effort required to learn to use the technology and is similar to the diffusion theoretic concept of complexity. Perceived usefulness refers to the user's subjective judgment of the utility of the prospective technology in increasing his or her job performance within the organization; it closely parallels the relative advantage factor from diffusion theory. Because complex technologies with steep learning curves reduce the perceived utility of the technology, in TAM, perceived ease of use influences usefulness. Together, these two instrumental beliefs mediate the influence of external factors on the user's intent to adopt the technology, which, in turn, predicts actual adoption.

In communication science, Vishwanath and Goldhaber (2003) presented a hybrid model that combined the TAM-based instrumental factors along with the diffusion theoretic factors such as observability and compatibility. Also included were factors such as change agent contact, media ownership, and media use, which mediated the influence of external demographic factors on the innovation specific beliefs; the innovation specific beliefs then directly influenced behavioral intent.

There were many subsequent modifications and additions to TAM from within the MIS field that culminated in the Unified Theory of Acceptance and Use of Technology (UTAUT) (Venkatesh et al., 2003). According to UTAUT, four beliefs about an innovation are core determinants of an individual's intent to adopt: performance expectancy, effort expectancy, social influence, and facilitating conditions. Performance expectancy and effort expectancy are similar to the innovation-diffusion constructs of relative advantage and perceived complexity, respectively; social influence is similar to the construct of image, while facilitating conditions taps into the innovation-diffusion construct of compatibility. Of these, facilitating conditions is considered the least important because its influence is not significant in the presence of performance expectancy and effort expectancy. Factors such as gender, age, experience, and voluntariness of use are thought to mediate the influence of these four core determinants on behavioral intent.

This contemporary approach to diffusion modeling also has a number of advantages. First, the models are robust and parsimonious. They account for a large percentage of intention (close to 70%) while using just a handful of measures. This makes the model particularly popular and appealing from an applied standpoint. Second, the measures developed are readily available and easily adapted to varying study contexts. This makes it not only easy to apply these measures but also justify their use, the latter being an important expectation of reviewers and journal editors. Not surprisingly, TAM and UTAUT are the most widely used models of adoption. These models have been applied to the study of a range of technological innovations ranging from Personal Digital Assistants (PDA) in hospitals (Vishwanath et al., 2008) to the adoption of information communication technologies within households (Dwivedi, Williams & Venkatesh, 2008).

Despite their advantages, the traditional and contemporary approaches to the measurement and modeling of adoption suffer from at least three major limitations. First, the use of either approach involves a trade-off. In the traditional approach, scholars have to choose between methodological rigor and the ability to measure nuanced variations in adopters' intentions. In the contemporary approach, the trade-off is between a parsimonious, good fitting research model and an explanation that is anything other than the obvious. Consider the case of a policymaker who utilizes TAM to explain to an organization that the technology the organization implements should be easy to use and useful. These instrumental features are but the most obvious minimum requirements that users would expect from anything that they adopt, be that new software, a complex innovation, or a coffee pot.

A second limitation of the modeling approaches is that these models assume that the same sets of finite, unidimensional valuations of potential innovations apply regardless of when, where, and by whom an innovation is potentially adopted. In other words, regardless of whether you are in Poughkeepsie or Panama, only perceived ease of use and usefulness would matter to you as an adopter. Empirical evidence, however, suggests otherwise. Within the MIS realm, studies that have tracked user preferences have found that users use different valuations prior to and after implementation (Karahanna, Straub & Chervany, 1999). Likewise, research in communication has shown that early adopters and later adopters cognitively conceptualize technologies they adopt differently (Vishwanath & Chen, 2006). Hence, it is very likely that different types of adopters use different dimensions of valuation to understand an innovation. This might be the case for employees within organizations as well as for the average individual consumer. For instance, upper management when evaluating a technology might consider price as the determinant of its usefulness; students might value social influence and image instead of ease of use; or later consumer adopters might value trialability over all other factors. Of course, these issues are potentially less significant among individuals who have little control over their decision to adopt (this would include prisoners and individuals in line functions in an organization), but for most individuals who are outside these constrained contexts, the extant diffusion models tend to underestimate the differences in valuations that ultimately influence their decision to adopt.

A final limitation is that all the different behavioral models of adoption focus on the potential adopters' decision to adopt a singular innovation that is often preselected for them by the researcher prior to model testing. That is, all the models are absolute models of intent rather than models of relative choice. In almost all human decision-making scenarios, however, individuals choose between multiple alternatives and utilize a personal valuation scheme to cognitively process and reduce the set of alternatives to one final choice (Simmons, 1974). Likewise, innovations exist in a spectrum of choice, and potential adopters utilize different valuation schemes to reduce the set of alternatives to one choice innovation that they eventually adopt. All the models of adoption ignore this fundamental aspect of adopter decision making.

Of course, one could argue that diffusion theoretic constructs of perceived compatibility and relative advantage capture relative valuations between alternative innovations. These constructs, however, measure the relationship between an innovation and one it supersedes. So the choice between competing innovations is clearly underestimated.

The primary reason for ignoring the choice processes in the manner in which choice is measured and modeled. While behavioral intent and beliefs are easily measured using metric measures, choice is always a categorical variable. Hence, choice between more than a few variables, while easy to measure, poses a challenge to modelers who use standard regression-based approaches that require metric data. Multidimensional scaling techniques provide an alternative means for accurately measuring and modeling such choice processes. This, however, needs a fresh view of the adoption process that can accommodate these modeling techniques. In the next section, we present a relational model of adopter choice that presents such an approach.

A Relational Model of Adopter Choice

The relational model of adopter choice models the associational process by which potential adopters choose an innovation from among multiple alternatives. These associational processes are fundamental to human cognition: the human mind organizes all stimuli based on their conceptual similarity with other objects, such that objects that are similar are more closely associated than those that are dissimilar. Such a view of cognition exists since the time of Aristotle (Anderson, 1983) and is frequently used as a basis for social and cognitive science research (Barnett, 1988; Carley, 1986; Shah et al., 2004; Vishwanath & Chen, 2006). Hence, the associational view is ideal for structuring and representing the interrelationships between multiple innovations.

Barnett and Siegel (1988) extended this associational perspective to explain the diffusion of innovations within social systems. The associational perspective draws from older conceptualizations of the collective consciousness by Durkheim (1895) and views diffusion as a cognitive process and innovations as combinations of associative elements of thought among collections of individuals in a system. New innovations provide an opportunity for new linkages or rearrangements of elements that have not been previously associated. As an innovation spreads though a social system, the conceptual configuration shared by the members change thereby altering their associative structures. The degree of reorganization of the associative elements is a function of the amount of information members of that system receive about the innovation (Woelfel & Barnett, 1982). Innovations that cause a minimal change in the associational structure tend to be closely related. Hence, the resultant associations indicate the compatibility with the values, past experiences, and needs of the potential adopter.

Our relational model of adopter choice extends this associational view to the study of multiple innovations, their relative associations with regard to each other and the individual adopter. These relationships are structured and presented as associations on a space that is empirically derived to reflect the preference of each adopting segment. The dimensions in the space reflect the valuation schemes of the adopting segment. Further, our relational model draws on the balance of forces or equilibrium concept of choice (Woelfel, 1975) and locates an ideal point on the space. This ideal point is essentially the average of the coordinate scores for an axis across all the dimensions on that space. Innovations that are most likely to be adopted are those that are closest to this ideal point. The relationships between innovations are measured by comparing the distance between innovations and the ideal point in the dimensional space.

Figure 2 presents a simple example of our relational model on a two-dimensional space. While the space has been represented on two dimensions for simplicity, it can be represented in as many dimensions as there are concepts in that space. In the example, the *x* axis represents "price" and the *y* axis represents "observability." Since the dimensions are empirically generated, differences in valuation schemes can be easily ascertained and mapped. For instance, if early adopters place a higher value on price and later adopters instead value trialability, these could be empirically determined and represented. There are five innovations represented by dashed vectors in the figure: BlackBerry (4,4), Sony PlayStation 3 (PS3) (6,2), and Amazon's Kindle (-2, 7). The equilibrium concept would locate the ideal point, closest to (4,4,3), and in this simple example, iPhone lies closest to the point, denoted by the solid vector. Hence, other factors being equal, the potential adopter would choose an iPhone; moreover, the adopter would compare iPhones with BlackBerrys and contrast them using price and observability.

The simple model could be expanded to accommodate any number of innovations along with a host of preferred attributes. Concepts representing individual attributes could be derived from the literature or elicited from the study population using projective techniques. The data to test the model could be empirically derived using paired comparisons that would provide a metric estimate of relationship among the innovations. Our relational model, thus, provides a quantitative approach to the study of adopters' relative choice among innovations: it accounts for the relationship among innovations, the relative preferences among them, as well the evaluative schema used to arrive at those preferences. The next section presents the methodology that is ideal suited to test this model.

Methodology

The test of the relational model requires a measurement system that meets the following requirements (Barnett, 1988; Barnett & Siegel, 1988; Vishwanath & Chen, 2006): (1) the links among a set of elements must be measured in terms of the similarity (or dissimilarity) among the elements. The measurement system needs to be capable of relating technologies to each other; (2) the measurement system needs to be holistic and capable of simultaneously measuring all integrated dimensions and produce a complete description of the relationships between technologies within and across different concepts; (3) it must involve consensual measures that allow for predictions at both the individual and the aggregated or system level; and (4) in order to make possible calculations of the rate of change and/or degree of differences between associations of adopters and non-adopters, the measurement scheme must ideally be at the ratio level.

The Galileo system satisfies all these demands (Woelfel & Fink, 1980). The Galileo system is an integrated methodology for metric multidimensional scaling using paired distance judgment data (Woelfel & Fink, 1980). Rice and Rogers (1984) recommend its use to describe the changes in conceptual structures of adopters of information technologies. Beninson-Germano et al. (1988) used the Galileo system to examine the adoption of automated ordering systems in the pharmaceutical industry.

Galileo measurement begins by measuring dissimilarities among a set of concepts describing a domain, defined by these concepts, using paired comparison (every possible pair of the concepts' dissimilarities are measured). So given N concepts, the dissimilarity among $N(N-1)/2$ pairs of symbols is estimated by respondents. In Galileo measurement, the "self" or "me" is also included as a concept since it provides a reference frame to understand the position of ego relative to the other concepts. Respondents are provided with a known distance estimate and instructed to judge the distances between pairs of concepts in proportion to this standard distance. The instruction is generally of the form, "If the distance between x and y are u units, how far apart are concepts a and b?"

The empirical distance data collected yields a matrix of concept dissimilarities. This matrix is transformed into a scalar products matrix that is double-centered to establish the origin at the centroid of the distribution. The matrix is then decomposed to achieve a coordinate matrix whose columns are orthogonal axes (dimensions) and rows are the projections of the concepts on the dimensions (Woelfel & Fink, 1980). Since the coordinates for the items represent the least-square balance points for the items relative to each other, the distances between objects (their coordinates) can be taken to indicate the conceptual similarity between the items. Thus, the greater the reported distance, the greater the conceptual differentiation, and vice versa.

In the relational innovation choice model, the innovation that is closest to the "self" in the collective space would be the innovation that is most preferred. In other words, innovations that are preferred are those that are maximally close to the individual, self point, i.e., have a distance of 0 from the self.

Hence, an adopter's likelihood of choosing an innovation from among other innovations can be predicted as follows:

$$P(x) = 1 - \frac{Sx}{Sx + Sy + Sz} \quad \text{[Equation 1]}$$

Where "P(x)" represents the probability of choosing innovation x, and Sx, Sy, Sz are the aggregate distances between the self and innovation's x, y, z, respectively, across all dimensions.

The Galileo system's multidimensional scaling method is considerably more precise when compared with traditional scale measurements. Gillham and Woelfel (1977) experimentally measured the precision of a Galileo-type pair-comparison instrument against other scaling methods. They found that even in extreme cases, when the Galileo ratio procedures might be only 50% reliable, it still generated about 2.70 times more reliable information when compared with a 10-point semantic differential scale, even if the semantic differential scales contained no random error (Woelfel & Fink, 1980). A vast body of communication, linguistic, and cross-cultural research has demonstrated the reliability and theoretical validity of the Galileo system (Barnett & Woelfel, 1988; Kincaid et al., 1983; Vishwanath & Chen, 2006, 2008). In the next section we present an empirical test of our relational model of adopter choice using the Galileo system.

Empirical Test of the Relational Model of Adopter Choice

Data and Operations

To test the relational model, data were gathered from undergraduate communication 101 students at the University of Buffalo. Subjects responded to a web-based questionnaire that contained the Galileo measures. The instrument asked respondents to consider the distance between internet and television to be 50 units. Using this referent, respondents were asked to estimate the distances among 105 concept pairs (15 concepts including "myself"). Table 1 presents a list of concepts used in the study.

Wherever possible, primitive concept terms were generated to reflect diffusion theoretic attributes of innovations. For instance, perceived complexity was represented by the term "easy to use," technological efficacy was represented by the term "ability to use," social influence was represented by family and friends. Constructs such as observability, trialability, usefulness, and purchase were represented as concepts by simply using their construct names. Three innovations were included: BlackBerry, PlayStation 3 (PS3), and iPhone.

The data for the study were collected November 14–21, 2007. At the time of the data collection, the iPhone was a relatively new innovation, and it was less than five months since it had been launched in the U.S market. The data collection netted 349 usable responses (57% female). In addition to measuring the Galileo data, respondents' technological innovativeness was measured using the six-item Domain-Specific Innovativeness scale (Goldsmith & Hofacker, 1991). Sample items in the measure read, "I am generally the first among my friends to purchase a new technological product," "Compared to most of my friends, I own more technological products," and so on. Items were measured using a 1–5 response scale (1 = strongly agree). The scale netted an alpha reliability of 0.91 (*mean* = *2.91*, *sd* = *0.88*). Respondents scoring in the first quartile of the innovativeness scale *(mean values less than 2.34)* were treated as early adopters, and the rest of the respondents were treated as later adopters. The mean distance matrix for early adopters is presented in Table 2 and for later adopters, in Table 3.

Results

Early adopter space. For early adopters, the scaling procedure resulted in 15 dimensions (13 real dimensions). Figure 3 presents the early adopters' space. Table 4 presents the spatial coordinate matrix for early adopters. The results report the findings for the first three dimensions because they are more easily visualized and because they account for close to half the variance in the data.

The first dimension is anchored by the attributes Affordability and Ease of Use, with the self concept leaning towards affordability. On this dimension, the technology closest to the self concept was BlackBerry (2.4 units), followed by iPhone (14 units), followed by PS3 (18.62). In 2007 when these data were collected, within the U.S., a BlackBerry could be purchased for under $100 (at no cost with some cell phone plans), a new iPhone cost $399, and PS3 cost $599.

The second dimension is anchored by the attributes Purchase and Affordability, with the self leaning towards purchase. On this dimension, the technology closest to the self concept was iPhone (4.9 units) followed by PS3 (9.36 units) and BlackBerry (14.34 units). This suggests that early adopters in our student sample are most likely to purchase an iPhone or a PS3. BlackBerry, positioned for business users, is the least likely to be purchased.

The third dimension is anchored by Observability and Affordability, with the self leaning towards observability. On this dimension, the self concept was closest to iPhone (1.98), followed by PS3 (2.59) and BlackBerry (6.11). Hence, for early adopters the iPhone would have the highest observability or social cachet followed by the PS3. In contrast, a BlackBerry would have relatively low social value.

Later adopter space. For later adopters, the scaling procedure again resulted in 15 dimensions (13 real). Figure 4 presents the late adopters' space. Table 5 presents the spatial coordinate matrix for later adopters. The results report the findings for the first three dimensions because they are more easily visualized and they account for close to half the variance in the data.

For later adopters, the first dimension is anchored by the attributes Affordability and Ease of Use; again the self leaned towards affordability. On this dimension, the technology closest to the self concept was BlackBerry (7.23 units), followed by iPhone (16.39 units), followed by PS3 (18.81). This finding paralleled the early adopter results.

The second dimension is anchored by the attributes Friends and Family and Affordability; the self concept leaned towards friends and family or social influence. On this dimension, the technology closest to the self concept was iPhone (5.47 units) followed by PS3 (9.26 units) and BlackBerry (13.37 units). This suggests that social influence is an important driver for later adopters while in contrast, for earlier adopters, vicarious consumption was more important.

The third dimension is anchored by Observability and Purchase with the self concept leaning towards observability. On this dimension, the self concept was closest to iPhone (4.25), followed by BlackBerry (4.97) and PS3 (5.53 units). Hence, for later adopters an iPhone followed by a BlackBerry would have relatively high observability, while a PS3 would have a relatively low social value.

Comparing early adopters and later adopter spaces. The comparison of early and later adopter spaces began with an evaluation of the overall spread of concepts within each space. This was done using the trace of the coordinates, the sum of the overall eigenvalues for each eigenvector matrix. For early adopters, the trace was 3,590, and for later adopters the overall sum was 4,634.

Next, the concepts that differentiated early and later adopters the most were estimated using the angles between the concept vectors across all the dimensions. Since the cosine of the angle between vectors reflects the correlation between them, smaller angles signify that the vectors are highly correlated or moving in the same direction. Table 6 presents the angles and correlations between the concept vectors for early and later adopters. Actual use (39.5 degrees), easy to use (30.9 degrees), and compatibility with existing technologies (30.5 degrees) had the lowest correlations. Friends and family (14.8 degrees), self (15.7 degrees), and ability to use (16.9) had the highest correlations and deviated the least.

A similar comparison of the angles between the first three dimensions for early and later adopters confirms that the two adopting segments utilize different sets of evaluative dimensions: the first dimension correlated the most (angle 12.6 degrees), followed by the third dimension (17.38 degrees); the second dimension deviated the least (24.5 degrees).

Finally, the likelihood of choosing each innovation was computed using equation 1 and using the coordinate distances between the innovations and the self on first three dimensions. Based on this, the early adopters in our sample had a 72% likelihood of choosing an iPhone while later adopters had a 65% likelihood of choosing an iPhone. While both segments had an equal likelihood (around 70%) of choosing a PS3, later adopters had a slightly higher (62%) likelihood of choosing the older innovation, the BlackBerry, over early adopters (56%).

Discussion

The empirical test of the relational model in many ways converges with the diffusion theoretical expectations for early and later adopters. Early adopters in our sample were much closer to actual purchase or adoption than later adopters. Moreover, the overall early adopter conceptual space was denser and less varied compared with later adopters, potentially because early adopters have greater awareness and knowledge of new technological innovations. Consistent with user acceptance research, the innovation's ease of use was an important consideration for early and later adopters. Furthermore, concepts such as actual use, easy to use, and compatibility significantly varied between early and later adopters. Differences in actual use and ease of use potentially reflect early and later adopters' personality-based differences in innovativeness and technological efficacy. The variances in the compatibility concept potentially reflect differences in how early and later adopters conceptualize the relationships between various technologies, an aspect that was empirically demonstrated by Vishwanath and Chen (2006).

The relational model, however, diverges from diffusion and user acceptance theories. The relational model demonstrated that early and later adopters utilized different sets of evaluative dimensions while considering innovations. Early adopters in the study weighed affordability against ease of use, purchase intent against affordability, and observability, i.e., the image or social cachet of the innovation against affordability. In contrast, earlier adopters weighed affordability against ease of use, friends and family members, i.e., what others in the social network thought about the innovation against affordability, and observability against purchase intent. Interestingly, the user acceptance factor of perceived usefulness was not a significant factor. Nor were diffusion theoretic factors such as compatibility or relative advantage. Instead the economic evaluation of the innovation's affordability was recurrently important and central to early and later adopters. Besides this, observability or image, social influence, and ease of use were the factors considered by early and later adopters of innovation.

These conclusions should be interpreted in light of the following limitations. First, the study utilized a student sample for the empirical data. This was done primarily for convenience and because students provided us with homogenous, controlled samples. Students, however, tend to be skewed in favor of adopting innovations, an issue that could account for some of our results. Additionally, the survey utilized brand names of technologies as concepts. This could have induced respondents towards considering issues such as observability and friends and family. The use of brand names could have also influenced individual consideration of price and affordability. Finally, the analysis relied on just the first three spatial dimensions and their coordinates. A more accurate understanding of adopter decision making would be netted by considering all the dimensions in the associational space as well as the influence of external factors such as Apple's advertising and of interpersonal influences from other adopters. The findings of the model are therefore preliminary, and additional research is necessary to validate and extend our findings. Future research using innovations as concepts (e.g., cell phone, home phone, and pagers), representative samples of adults, and more attributes could easily validate the model.

The relational model of adopter choice, however, provides a fresh approach to the study of adopter decision making. The model and the methodology makes it possible to study the absolute influence of an innovation's attributes on adopter intent while simultaneously accounting for the relative associations between the different innovations considered during this decision-making process. Similar to traditional diffusion models, the relational model can be applied in an exploratory manner, but it is more dynamic because it can account for the process by which adopters choose between any number of innovations and their evaluations of these innovations. Also, similar to contemporary intention models, the relational model can be used to confirm the evaluative dimensions found across different studies. Further, because the approach relies on simple, primitive definitions of concepts, the measures can be easily extended and compared against other domains of study in virtually any social system. Finally, since the model provides predictive probabilities of adopters' final choice, the veracity of these predictions can be easily confirmed on a posteriori basis. Thus, by accounting for the choice between different innovations, the relational model provides a more comprehensive understanding of the adopter decision-making process. This aspect of innovation decision making is the least well understood and, in many ways, one of the most important facets of the adopter decision-making process.

Tables and Figures

Table 1. List of concepts used in the study.

Useful
Easy to use
Friends and family
Compatible with existing technology
Observable
Trialable
Purchase
Use
Ability to use
Apprehension
BlackBerry
iPhone
PS3
Myself
Affordable

Table 2. Mean Distance between Concepts among Early Adopters.

CONCEPTS	1	2	3	4	5	6	7	8	9	10	11	12	13	14	15
1. Useful	0.00														
2. Easy to use	18.52	0.00													
3. Friends and family	17.81	24.31	0.00												
4. Compatible	17.15	18.67	23.79	0.00											
5. Observable	20.12	19.33	21.38	19.46	0.00										
6. Trialable	21.52	23.40	22.50	21.10	25.17	0.00									
7. Purchase	20.88	20.48	24.12	21.44	23.15	20.50	0.00								
8. Use	18.38	16.88	21.13	23.58	18.77	19.02	21.79	0.00							
9. Ability	20.65	18.29	21.65	24.19	21.44	19.38	23.79	20.77	0.00						
10. Apprehension	22.19	22.40	21.56	23.33	22.33	25.33	23.62	19.56	26.23	0.00					
11. Blackberry	21.83	19.35	23.29	22.19	21.15	22.79	28.88	23.23	20.54	25.31	0.00				
12. iPhone	23.23	20.23	24.87	20.54	21.38	19.48	26.21	21.21	19.02	20.65	24.21	0.00			
13. PS3	27.85	21.15	27.77	20.73	21.31	22.44	25.87	22.62	20.58	24.00	25.44	23.42	0.00		
14. Myself	20.42	21.87	17.81	19.08	19.58	20.69	23.25	19.96	21.56	23.96	23.48	24.58	26.21	0.00	
15. Affordable	26.50	26.83	25.73	25.56	22.79	23.85	27.50	26.58	23.60	25.83	23.73	29.38	29.56	26.31	0.00

Table 3. Mean Distance between Concepts among Later Adopters.

CONCEPTS	1	2	3	4	5	6	7	8	9	10	11	12	13	14	15
1. Useful	0														
2. Easy to use	20.52	0													
3. Friends and family	25.45	27.73	0												
4. Compatible	22.54	24.22	26.82	0											
5. Observable	27	26.55	25.38	25.25	0										
6. Trialable	25.54	24.63	29.97	26	24.78	0									
7. Purchase	27.77	28.56	30.95	25.88	27.51	24.88	0								
8. Use	17.73	20.22	26.42	24.31	26.41	21.88	21.04	0							
9. Ability	18.67	19.76	25.51	22.53	25.61	23.79	24.97	19.08	0						
10. Apprehension	31.22	26.55	27.95	25.65	25.5	25.44	28.86	25.72	28.89	0					
11. BlackBerry	23.77	23.68	31.58	20.52	24.75	25.33	26.81	23.32	22.77	26.17	0				
12. iPhone	22.87	22.16	32.8	19.42	25.01	25.01	28.5	23.37	23.17	25.22	22.33	0			
13. PS3	29.86	23.04	29.61	20.46	23.91	24.85	27.09	25.81	23.55	26.22	30.23	31.11	0		
14. Myself	20.55	23.16	21.56	22.84	21.61	25.09	25.95	23.26	23.15	23.77	28.15	28.44	28.68	0	
15. Affordable	29.19	31.58	28	28.56	27.38	27.62	24.16	26.74	26.23	27.08	29.54	34.39	32	25.33	0

Table 4. Spatial Coordinates for Early Adopters on the First Eight Attribute Dimensions.

	Dimension 1	Dimension 2	Dimension 3	Dimension 4	Dimension 5	Dimension 6	Dimension 7	Dimension 8
Useful	-4.93	-6.59	-2.96	-4.29	3.16	3.38	4.20	-0.38
Easy to use	5.25	-0.30	-3.44	-1.64	5.21	7.96	0.24	0.39
Friends and family	-9.32	-4.93	-3.11	-2.81	-5.12	-3.65	-4.00	0.35
Compatibility	2.38	-1.80	-0.13	0.82	9.49	-7.05	6.58	0.28
Observability	-0.68	1.83	-5.49	4.38	4.50	-1.04	-4.96	-6.49
Trialability	0.91	0.95	10.69	-4.95	-4.92	-2.85	3.11	4.08
Purchase	0.26	-9.50	10.40	3.41	4.40	5.02	-0.06	-0.85
Use	1.66	-4.19	-1.48	0.02	-4.44	5.54	-4.65	3.03
Ability to use	2.10	6.36	2.88	-7.19	-3.08	3.56	-3.74	-5.09
Apprehension	0.72	-4.65	-5.36	11.38	-7.55	-0.01	2.88	3.04
BlackBerry	-3.02	9.96	-6.69	-4.73	2.24	2.09	2.96	6.87
iPhone	8.73	0.53	-2.57	-2.27	-7.42	-2.35	7.42	-6.71
PS3	13.14	4.99	2.01	3.90	1.82	-4.37	-6.74	3.68
Myself	-5.49	-4.38	-0.59	-4.29	1.84	-7.56	-5.06	-0.09
Affordable	-11.69	11.71	5.85	8.25	-0.11	1.35	1.81	-2.11

Table 5. Spatial Coordinates for Later Adopters on the First Eight Attribute Dimensions.

	Dimension 1	Dimension 2	Dimension 3	Dimension 4	Dimension 5	Dimension 6	Dimension 7	Dimension 8
	-3.98	-4.64	-2.44	-9.68	3.14	3.70	3.10	-1.81
Useful	5.99	-0.55	-4.05	-5.15	2.59	6.93	-1.90	1.53
Easy to use	-11.49	-6.48	-3.72	-1.35	-3.90	-5.25	-7.45	-1.82
Friends and family	4.40	0.89	-1.65	-0.23	6.64	-8.64	6.68	1.29
Compatibility	-0.10	3.83	-5.75	4.81	2.27	-4.44	-7.15	-6.47
Observability	2.55	4.27	8.32	-3.98	-7.81	-2.55	0.50	5.26
Trialability	0.17	-5.48	13.59	5.03	4.32	4.46	1.81	-1.27
Purchase	0.87	-5.59	1.76	-1.40	-2.89	9.36	-0.76	1.89
Use	1.01	4.73	1.96	-7.57	-1.86	4.11	-3.06	-5.13
Ability to use	1.31	-1.38	-5.53	12.27	-9.23	-3.39	1.99	5.83
Apprehension	-0.15	7.16	-6.70	-2.83	3.68	4.81	7.37	6.35
BlackBerry	9.21	-0.75	-5.98	-2.75	-3.42	0.20	12.22	-7.24
iPhone	11.73	3.04	3.80	4.91	3.30	-4.33	-9.97	4.16
PS3	-7.09	-6.22	-1.73	-2.11	2.54	-6.65	-4.53	-0.52
Myself	-14.43	7.16	8.13	10.02	0.63	1.69	1.15	-2.06

Table 6. Relationship between Concepts between the Early and Later Adopter Spaces.

Concept	Early adopters' magnitude	Later adopters' magnitude	Scalar products	Vector correlation	Angle
1	13.75	16.30	203.40	0.91	24.90
2	13.12	16.10	181.41	0.86	30.90
3	15.75	20.54	312.74	0.97	14.80
4	14.02	15.22	183.84	0.86	30.50
5	13.58	17.29	217.50	0.93	22.10
6	14.62	17.13	220.39	0.88	28.30
7	17.03	18.93	299.87	0.93	21.50
8	13.28	14.22	145.80	0.77	39.50
9	14.08	14.47	195.75	0.96	16.10
10	16.46	19.00	288.09	0.92	23.00
11	16.42	17.71	258.56	0.89	27.30
12	15.84	18.39	264.64	0.91	24.70
13	17.75	19.40	329.86	0.96	16.70
14	14.85	15.89	227.28	0.96	15.70
15	19.86	21.24	401.76	0.95	17.70

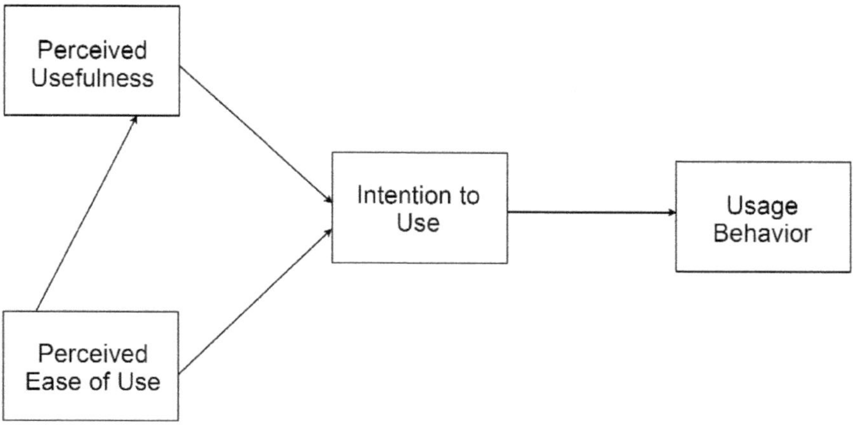

Figure 1. The Technology Acceptance Model (TAM).

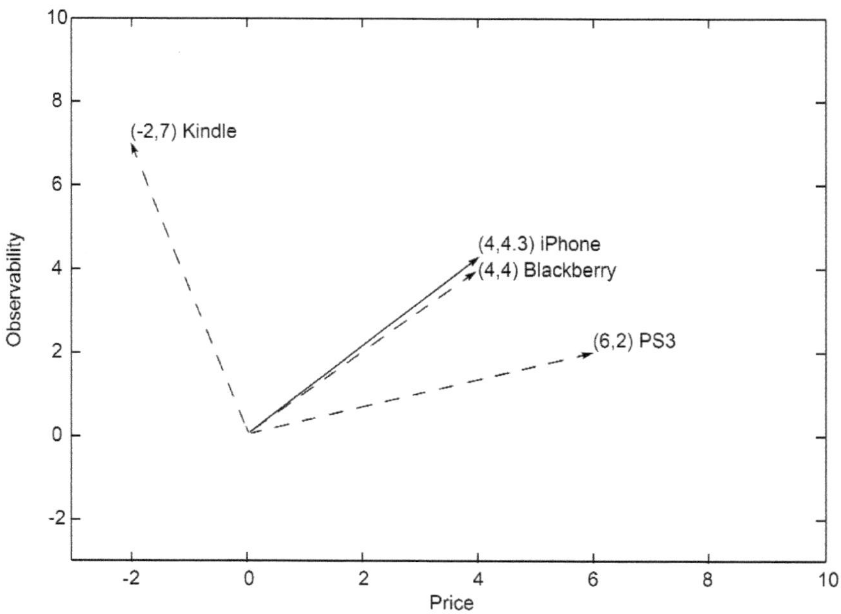

Figure 2. Two-dimensional Representation of the Relational Model of Adopter Choice.

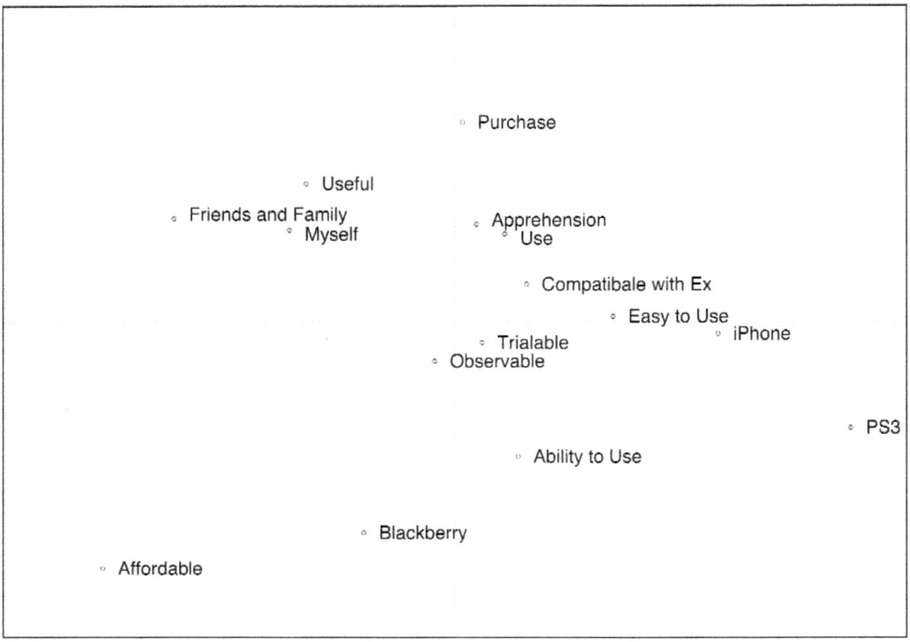

Figure 3. Early Adopters' Space.

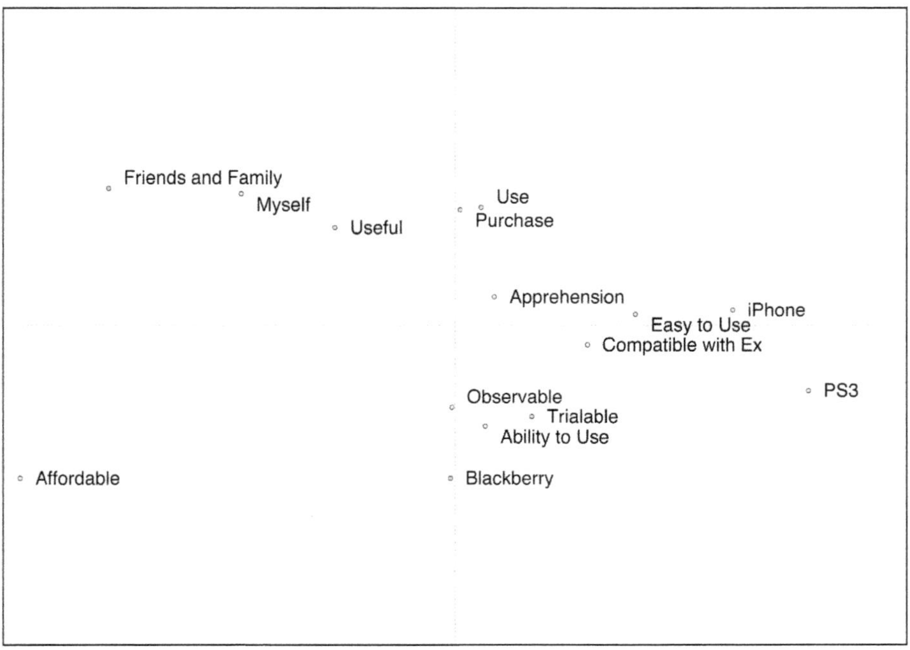

Figure 4. Later Adopters' Space.

References

Anderson, J. R. (1983). *The architecture of cognition*. Cambridge, MA: Harvard University Press.

Barnett, G. A. (1988). An associational model for the diffusion of complex innovations. In G.A. Barnett & J. Woelfel (Eds.), *Readings in the Galileo System Theory, Methods and Applications* (pp. 55–74). Dubuque, IA: Kendall/Hunt.

Barnett, G. A. & Siegel, G. (1988). The diffusion of computer assisted legal research systems. *Journal of the American Society for Information Science, 39*, 224–234.

Barnett, G. A. & Woelfel, J. (Eds.). (1988). *Readings in the Galileo System*. Dubuque, IA: Kendall/Hunt.

Beninson-Germano, M., Barnett, G. A., Dinkelacker, J. W., & Downing, K. W. (1988). Automated ordering in the pharmaceutical industry. In G.A. Barnett & J. Woelfel (Eds.), *Readings in the Galileo System* (pp.265–294). Dubuque, IA: Kendall/Hunt.

Carley, K. A. (1986). An approach for relating social structure to cognitive structure. *Journal of Mathematical Sociology, 12*, 137–189.

Davis, F. D., Bagozzi, R. P. & Warshaw, P. R. (1989). User acceptance of computer technology: A comparison of two theoretical models. *Management Science, 35*, 982–1003.

Durkheim, E. (1895). *The division of labor in society*. New York, NY: Free Press.

Dwivedi, Y. K., Williams, M. D., & Venkatesh, V. (2008). A profile of adoption of information & communication technologies (ICT) research in the household context. *Information Systems Frontiers, 10*, 385–90.

Fishbein, M., & Ajzen, I. (1975). *Belief, attitude, intention, and behavior: An introduction to theory and research*. Reading, MA: Addison-Wesley.

Gillham, J., & Woelfel, J. (1977). The Galileo system of measurement. *Human Communication Research, 3*, 223–234.

Goldsmith, R. E., & Hofacker, C. F. (1991). Measuring consumer innovativeness. *Journal of the Academy of Marketing Science, 19*, 209–221.

Karahanna, E., Straub, D. W., & Chervany, N. L. (1999). Information technology adoption across time: A cross-sectional comparison of pre-adoption and post-adoption beliefs. *MIS Quarterly, 23*, 183–213.

Kincaid, D. L., Yum, J. O., Woelfel, J., & Barnett, G. A. (1983). The cultural convergence of Korean immigrants in Hawaii: An empirical test of a mathematical theory. *Quality and Quantity, 18*, 59–78.

Leung, L., & Wei, R. (1999). Who are the mobile phone have-nots? Influences and consequences. *New Media & Society, 1*, 209–226.

Rice, R. E., & Rogers, E. M. (1984). New methods and data for the study of new media. In R. E. Rice (Ed.), *The new media: Communication, research, and technology* (pp. 81–99). Beverly Hills, CA: Sage.

Rogers, E. M. (2003). *Diffusion of innovations* (5th ed.). New York: Free Press.

Shah, D. V., Kwak, N., Schmierbach, M., & Zubric, J. (2004). The interplay of news frames on cognitive complexity. *Human Communication Research, 30*, 102–120.

Simmons, P. J. (1974). *Choice and demand*. New York: John Wiley.

Venkatesh, V., Morris, M. G., Davis, G. B., & Davis, F. D. (2003). User acceptance of information technology: Toward a unified view. *MIS Quarterly, 27*, 425–478.

Vishwanath, A., Brodsky, L., Shaha, S., Leonard, M., & Cimino, M. (2009). Prescriber attitudes toward PDA prescription-assistance technology. *International Journal of Medical Informatics*, *78*, 330–339.

Vishwanath, A., & Chen, H. (2006). Technology clusters: Using multidimensional scaling to evaluate and structure technology clusters. *Journal of the American Society for Information Science & Technology*, *57*, 1451–1460.

Vishwanath, A., & Chen, H. (2008). Personal communication technologies as an extension of the self: A cross-cultural comparison of people's associations with technology and their symbolic proximity with others. *The Journal of the American Society for Information Science & Technology*, 59, 1761–1775.

Vishwanath, A., & Goldhaber, G. M. (2003). An examination of the factors contributing to adoption decisions among late diffused technology products. *New Media & Society*, 5, 547–572.

Woelfel, J. (1975). Theoretical issues and alternatives: A theory of occupational choice. In J. S. Picou & R. E. Campbell (Eds.), *Career behavior of special groups* (pp. 41–61). Columbus, Ohio: Merrill.

Woelfel, J. & Barnett, G. A. (1982). Multidimensional scaling in Riemann Space. *Quality and Quantity: A European Journal of Methodology*, *16*, 469–491.

Woelfel, J. & Fink, E. L. (1980). *The measurement of communication processes: Galileo theory and method*. New York: Academic Press.

· 2 ·

Three Theoretical Perspectives on Communication Technology Adoption

Oscar Peters

Introduction

In this chapter, three generally accepted models in the field of communication science to explain and predict communication technology adoption behavior will be discussed and compared. First, the expectancy-value perspective on uses and gratifications will be introduced, followed by a social cognitive perspective on communication technology adoption. The third perspective is the unified theory of acceptance and use of technology. The convergence between the three theoretical perspectives on the central processes and phenomena related to the understanding of communication technology adoption will be discussed, and the differences in performance of the three models to explain and predict communication technology adoption behavior will be compared. The most important implication that can be derived from this chapter is that the choice of a model to understand a particular communication technology adoption behavior should be determined foremost by the stage of development and diffusion the particular communication technology is in. Depending on whether the communication technology in question is an already well-accepted communication technology or a new innovative communication technology, an appropriate communication technology adoption model should be selected to respectively explain and predict communication technology adoption behavior.

In the year 1954, in one of the first mass communication textbooks, *The Process and Effects of Mass Communication*, Wilbur Schramm poses the question, "What determines which offerings of mass communication will be selected by a given individual?" (p. 19). The answer to this question is what Schramm calls the "fraction of selection." People weigh the level of reward they expect from a given medium or message against how much effort they must make to secure that reward. According to Schramm, individuals make media and content choices based on the expectation of reward and effort required.

Schramm's question in 1954 is still relevant today, perhaps even more relevant as the offers, choices, and possibilities of new communication technologies are more diverse and dynamic than in the 1950s. In this chapter, three generally accepted models in the field of communication science will be discussed and compared. First, the expectancy-value perspective on uses and gratifications will be introduced, followed by a social cognitive perspective on communication technology adoption. The third perspective is the unified theory of acceptance and use of technology. At the end of the chapter, the convergence between the three perspectives on the central processes and phenomena related to the understanding of communication technology adoption will be discussed. The aim of this chapter is not to evaluate and compare the three perspectives in terms of good or bad, weak or strong. Instead, the perspectives are evaluated on a continuum that ranges from "very useful" at one end to "not particularly useful" at the other end to either predict or explain communication technology adoption.

The Uses and Gratifications Approach and Expectancy-Value Theory

One of the first research approaches in the communication research tradition to focus on media use and adoption is the uses and gratifications approach. According to Bryant and Miron (2004), the year 1959 can be considered as the official birth of the uses and gratifications, when Bernard Berelson claimed that communication research appeared to be dead, and Elihu Katz responded that research should move from what media do to people to what people do with media. Infante, Rancer, and Womack (1997) posit the start of uses and gratifications approach with the first work on uses and gratifications published in 1944 by Lazarsfeld and Stanton on radio research.

At the core of the uses and gratifications approach lies the assumption that audience members actively seek out the mass media to satisfy individual needs.

Katz, Gurevitch, and Haas (1973) believed that audience members actively use various media to fulfill certain needs or goals. According to Katz et al., audience members choose a medium and allow themselves to be swayed, changed, and influenced—or not. Two other assumptions according to Infante et al. (1997) are that audiences also use media to fulfill expectations and that audience members are aware of and can state their own motives for using mass communication. According to Infante et al., communication theorists had three objectives in developing uses and gratifications research. First, they hoped to explain how individuals use mass communication to gratify their needs. Second, their objective was to discover the underlying motives for an individual's media use. And third, they wanted to identify positive and negative consequences of individual media use, such as parasocial interaction effects (e.g., TV viewing as a form of companionship), emotional effects (e.g., to see a movie for entertainment or to escape from everyday life), and behavioral effects (e.g., reading a newspaper to pass time or because it is a habit).

In a more general definition of uses and gratifications, Katz, Blumler, and Gurevitch (1974) posit that uses and gratifications research is "concerned with the social and psychological origins of needs, which generate expectations of the mass media or other sources, which lead to differential patterns of media exposure (or engagement in other activities), resulting in need gratifications and other consequences, perhaps mostly unintended ones" (p. 20).

The most central concept in the uses and gratifications research tradition is probably the concept of gratifications sought (Hendriks Vettehen, 1998, 2002). Despite the importance of the central concept of gratification, a generally accepted definition of the concept itself is not easy to find in the rich uses and gratifications literature. According to Ruggiero (2000), one of the continued flaws in uses and gratifications is that there still exists a lack of clarity among the central concepts and that uses and gratifications researchers attach different meanings to the concepts. Early uses and gratifications researchers were trying to explain media use by an inventory of the consequences of media use people experienced. These experienced gratifications were used to explain media use. Typical for these gratifications is that they are the result of media use. Here lies one of the main critics of the uses and gratifications approach; namely, that media use is explained by the consequences it has for the user. According to Hendriks Vettehen (1998), it seems that a circular argument is used: use leads to desired gratifications, but the desire to receive these gratifications is also the reason for use (p. 17).

A number of media scholars stressed the need to distinguish between the motives for media consumption and the gratifications perceived from this experience (e.g., Greenberg, 1974; Katz et al., 1974; Rosengren, 1974). With the distinction between gratifications sought and gratifications obtained there is no longer a circular argument to explain media use, because media-use motives no longer follow the evaluation of media use. As Hendriks Vettehen (1998) illustrates: "The evaluation that watching television on an evening brought some entertaining does not imply that the need for entertainment has been the motive to watch television" (p. 18). By the division of gratifications into two concepts it should be possible in theory to explain the changes in media use by the discrepancy between gratifications sought (motives) and gratifications obtained (evaluation).

The question is whether this theoretical difference between motives and evaluation can also be shown empirically. Hendriks Vettehen (1998, 2002) states that the analytic difference uses and gratifications researchers make between media-use motives and other relevant concepts are not yet empirically distinct. According to Hendriks Vettehen, an elaborated alternative to the measurement of motivations may be found in the expectancy-value approach, in particular the application of the measurement of $b_i e_i$ as determinant of media exposure (with b_i = belief that some object of exposure possesses attribute i and e_i = evaluation of attribute i).

To provide uses and gratifications with a more solid theoretical basis, several authors (e.g., Galloway & Meek, 1981; Rayburn & Palmgreen, 1984) moved away from the origin of needs perspective and incorporated an expectancy-value perspective as used within social psychology (e.g., Fishbein & Ajzen, 1975) into uses and gratifications research (see Figure 1).

Although the various theories under the label expectancy-value differ somewhat in their emphases, according to Palmgreen (1984) all view behavior, behavioral intentions, or attitudes (or all three) as a function of (1) expectancy (or belief)—the perceived probability that an object possesses a particular attribute or that a behavior will have a particular consequence; and (2) evaluation—the degree of affect, positive or negative, toward an attribute or behavioral outcome.

Fishbein and Ajzen's (1975) attitude-behavior model, known as the theory of reasoned action, probably constitutes the most influential and well-documented expectancy-value model of attitudes and decision-making (Aarts, Verplanken & Van Knippenberg, 1998). The theory of reasoned action postulates that attitudes (the desirability of the behavior, which is considered to be a function of the sum of the perceived values of the expected consequences of the behavior), together with subjective norms (representing the experienced

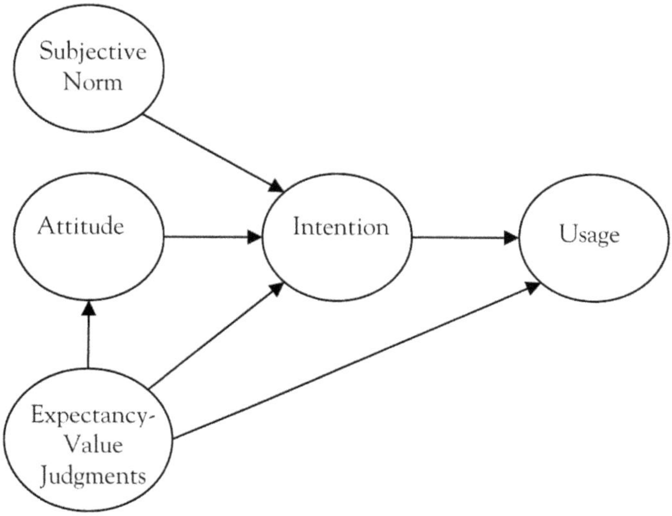

Figure 1. The Expectancy-Value Judgments Model of Uses and Gratifications.

social pressure), are the antecedents of behavioral intentions, which in turn are supposed to precede behavior. Because the attainment of behavioral goals is not always completely under volitional control, Ajzen (1991) has added a third concept to the prediction of behavior, perceived behavioral control, representing one's perception of how easy or difficult it is to perform the behavior. The inclusion of perceived behavioral control has resulted in the theory of planned behavior.

Infante et al. (1997) state there are two major explanations to the expectancy-value mechanism: affective-cognitive consistency (Rosenberg, 1956) and learning theory (Cronkhite, 1969). According to affective-cognitive consistency, people have affect and cognitions regarding a topic or object and try to make the two consistent. Affect involves attitude—how favorably people evaluate an object. Cognitions are beliefs about what is related to the object. Affective-cognitive consistency proposes a law of cognitive behavior: if you change a person's belief about a topic, object, or proposal, the attitude will "automatically" change in the same direction and to the same degree as the belief changes (Infante et al., 1997, p. 167). According to learning theory, the idea is that people learn to associate consequences with behavior. The response consequences (such as success or failure, or rewards or punishments) influence the likelihood that a person will perform a particular behavior again in a given situation (cf. Stone, 1998).

The expectancy-value perspective is a widely used theoretical approach in studying the adoption, usage, and consumption of mass media and also new communication technologies. For instance, Babrow and Swanson (1988) extended the application of expectancy-value theory to gratifications research in a study of student exposure to television news. Babrow (1989) used an expectancy-value approach to untangle student perceptions of soap opera viewing. Leung and Wei (1999) examined the use of the pager as a newly emerged mass medium for seeking news, focusing on the effects of expectancy-value judgments of the use of the pager in general and on the use of pagers as a news medium; in particular, on the level of exposure to news. More recently, Book and Barnett (2006) adopted an expectancy-value approach to examine the potential of PCTV (watching television on your PC) among consumers.

Social Cognitive Theory

According to LaRose, Mastro, and Eastin (2001, p. 397) the gratifications sought–gratifications obtained formulation as used by uses and gratifications researchers is "seemingly indistinguishable" from an important mechanism in social cognitive theory, i.e., enactive learning (Bandura, 1986). Enactive learning describes how humans learn from experience. In the social-cognitive view, interactions with the environment influence media exposure by continually reforming expectations about the likely outcomes of future media consumption behavior. Seemingly, this represents the same process that describes the relationship among gratifications sought, media behavior, and gratifications obtained (Palmgreen et al., 1985). Actually, according to LaRose et al., the outcome expectation construct parsimoniously bridges the gulf between gratifications sought and gratifications obtained in uses and gratifications research.

Within social cognitive theory, human behavior is defined as a triadic, dynamic, and reciprocal interaction of personal factors, behavior, and the environment (Bandura, 1986). This triadic causal mechanism is mediated by symbolizing capabilities that transform sensory experiences into cognitive models that guide actions. Within social cognitive theory, behavior is an observable act and the performance of behavior is determined, in large part, according to LaRose and Eastin (2004), by the expected outcomes of behavior, expectations formed by our own direct experience (enactive learning) or mediated by vicarious reinforcement observed through others (vicarious learning).

Outcome expectations, defined as judgments of the likely consequences of behavior (Bandura, 1997), provide incentives for enacting behavior, whereas expectations of aversive outcomes provide disincentives (Bandura, 1986). Expected outcomes are organized around six basic types of incentives for human behavior (Bandura, 1986, p. 232). These include monetary incentives, social incentives (such as obtaining approval from others), and status incentives. Sensory incentives involve exposure to pleasing or novel sensations. Preferences for enjoyable activities are the basis for activity incentives. There are also internal, self-reactive incentives resulting from comparisons of personal actions with standards for behavior. According to LaRose and Eastin (2004) these incentives are theoretically constructed rather than statistically derived from exploratory factor analysis.

Other concepts from social cognitive theory that are important to understand communication technology adoption behavior are self-efficacy and self-regulation. Self-efficacy is the belief in one's capability to organize and execute a particular course of action (Bandura, 1997). Those who perceive themselves to be highly efficacious with reference to a particular task will invest sufficient levels of effort to achieve successful outcomes, whereas those with low levels of self-efficacy will not persist. LaRose and Eastin (2004) posed that self-efficacy is directly related to media usage and indirectly related to media usage through expected outcomes. Prior experience, in turn, causally precedes self-efficacy (Eastin & LaRose, 2000), probably through the process of enactive mastery (Bandura, 1986). The social cognitive theory construct of self-regulation (Bandura, 1991) describes how individuals monitor their own behavior, judge it in relation to personal and social standards, and apply self-reactive incentives to moderate their behavior. Within social cognitive theory, habit is a failure of the self-monitoring subfunction of self-regulation. A related concept to habit is deficient self-regulation, a state in which conscious self-control is diminished (LaRose & Eastin, 2004).

Although habit and deficient self-regulation have not been clearly empirically distinguished in prior research, LaRose, Lin, and Eastin (2003) proposed a possible theoretical distinction, where habit represents the failure of self-monitoring and deficient self-regulation represents a failure of the judgmental and self-reactive subfunctions. According to LaRose and Eastin (2004), deficient self-regulation reflects a state of mind distinct from one in which media consumers are inattentive, explaining how both might have independent effects on media attendance. Habit strength and deficient self-regulation should be related by the fact that persons with deficient self-control may also be expected to engage in habitual behavior. Habit strength is expected to influence ongoing behavior. According to LaRose and Eastin, repetition makes us inattentive to the reasoning behind our media behavior; our mind no longer devotes attention resources to evaluating it, freeing itself for more important decisions. LaRose and Eastin posed that habit strength should be causally determined by outcome expectations, which precede habit strength in time. Habit strength should be preceded by self-efficacy, since users are unlikely to be inattentive to behavior they are still mastering.

The comprehensiveness and complexity of social cognitive theory make it somewhat difficult to operationalize, and many applications of social cognitive theory focus on one or two constructs, such as self-efficacy (e.g., Hofstetter, Zuniga & Dozier (2001), while ignoring the others (Stone, 1998). Although social cognitive theory is a broad theory of human behavior, it has also been applied to the context of communication technology adoption. See, for instance, Bandura's book chapter "Social Cognitive Theory of Mass Communication" in Bryant and Zillman (2001); the study by Hofstetter et al. (2001) to validate the concept of media self-efficacy in using television, newspaper, and interpersonal communication to monitor politics in everyday life; and LaRose, Lai, Lange, Love, and Wu's (2005) study on downloading behavior.

Inspired by Bandura's social cognitive theory, LaRose and Eastin (2004) proposed and tested a new model of media attendance that builds upon the conventional uses and gratifications approach (see Figure 2). According to LaRose and Eastin, the model of media attendance extends the uses and gratifications paradigm within the framework of social cognitive theory by instituting new operational measures of gratifications sought reconstructed as outcome expectations. According to LaRose et al. (2001, p. 399), attempts made by uses and gratifications researchers (e.g., Babrow & Swanson, 1988) to distinguish gratifications from formulations involving outcome expectations were of no avail and failed to produce more robust explanations of media exposure.

Unified Theory of Acceptance and Use of Technology

In the context of technology acceptance research, another perspective on the understanding of people's behavior for adopting and using communication technology is proposed: the multi-attribute model of technology use called the technology acceptance model (Davis, 1989). In principle, a multi-attribute

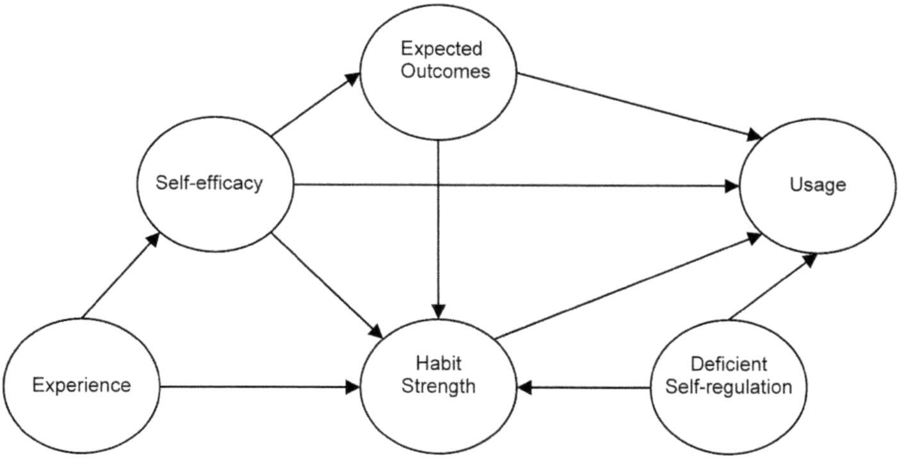

Figure 2. The Model of Media Attendance.

model represents a decomposition of a decision problem into smaller and less complex subproblems. The technology acceptance model examines how users come to accept and use a technology (Bagozzi, Davis & Warshaw, 1992; Davis, Bagozzi & Warshaw, 1989). According to Venkatesh et al. (2003), information systems research has long studied how and why individuals adopt new information technologies, and the explanation of user acceptance of new technology is often described as one of the most mature research areas in the contemporary information systems literature. Within this broad area of inquiry, there have been several streams of research, and each of these streams makes important and unique contributions to the literature on user acceptance of information technology (Venkatesh et al., p. 427). One stream of research focuses on individual acceptance of technology by using intention or usage as a dependent variable (e.g., Compeau & Higgens, 1995; Davis et al., 1989). Other streams have focused on implementation success at the organizational level (e.g., Leonard-Barton & Deschamps, 1988) and task-technology fit (e.g., Goodhue, 1995), among others.

Technology adoption is one of the most widely researched topics in information systems research. According to Venkatesh (2006), it has been studied at the individual (e.g. Venkatesh et al., 2003), group (e.g., Sambamurthy & Chin, 1994), and organizational (e.g., Fichman & Kemerer, 1997) levels. In terms of the reach, the technology acceptance model has been applied in a variety of domains, extending well beyond the initial scope of computer software studied by Davis (1989). According to Venkatesh, the technology acceptance model has been applied from marketing contexts (e.g., Dabholkar & Bagozzi, 2002; Gentry & Calantone, 2002; Yang & Peterson, 2004) to green electricity use (Arkesteijn & Oerlemans, 2005) to dairy farming (Flett, Alpass, Humphries, Massey, Morriss & Long, 2004).

The technology acceptance model replaces many of the attitude measures of the theory of reasoned action (Fishbein & Ajzen, 1975) with two technology acceptance measures: ease of use and usefulness. The technology acceptance model, which has strong behavioral elements, assumes that when someone forms an intention to act they will be free to act without limitation. In the real world there will be many constraints, such as limited ability, time, environmental or organizational limits, or unconscious habits that will limit the freedom to act (Bagozzi et al., 1992).

Venkatesh and Davis (2000) extended the original technology acceptance model to explain perceived usefulness and usage intentions in terms of social influence and cognitive instrumental processes. In an attempt to integrate the main competing user-acceptance models, Venkatesh et al. (2003) have formulated the unified theory of acceptance and use of technology. The models Venkatesh et al. reviewed and synthesized into the unified theory of acceptance and use of technology are the theory of reasoned action (Fishbein & Ajzen, 1975), the technology acceptance model (Davis, 1989), the motivational model (Davis, Bagozzi & Warshaw, 1992), the theory of planned behavior (Ajzen, 1991), a model combining the technology acceptance model, and the theory of planned behavior (Taylor & Todd, 1995), the model of PC utilization (Thompson, Higgins & Howell, 1991), the diffusion of innovations theory (Rogers, 2003), and the social cognitive theory (Bandura, 1986).

The unified theory of acceptance and use of technology comprises four core determinants of intention and usage and up to four moderates of key relations. According to Venkatesh, et al. (2003), four constructs play a significant role as direct determinants of user acceptance and usage behavior: performance expectancy, effort expectancy, social influence, and facilitating conditions. The four key moderators in the unified model of acceptance and use of technology are gender, age, experience, and voluntariness of use (see Figure 3). According to Venkatesh et al. (2003) in terms of explained variance the unified model is a substantial improvement over any of the original eight models and their extensions. Venkatesh et al. posed that the unified theory of acceptance and use of technology provides a useful tool to assess the likelihood of success for new technology introductions and helps to understand the drivers of acceptance in order to proactively design interventions targeted at populations of users that may be less inclined to adopt and use new systems.

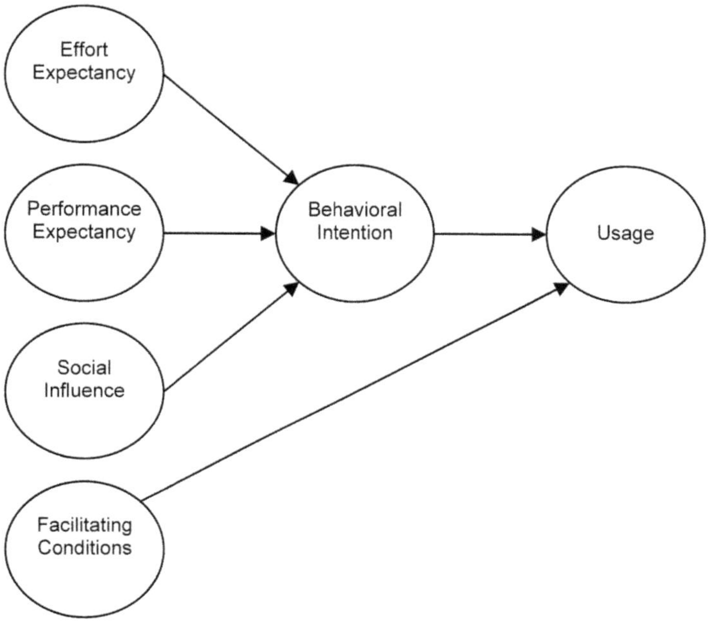

Figure 3. The Unified Model of Acceptance and Use of Technology.

Convergence on Central Processes and Phenomena Between the Three Theoretical Perspectives

The presented theoretical perspectives to understand communication technology use and adoption each cover broad bodies of literature with rich research traditions behind them, yet they also converge on the central processes and phenomena related to the formation of users' intentions to adopt communication technology, as part of an extended model of communication technology adoption behavior (cf. Stafford, Stafford & Schkade, 2004).

As already mentioned, Fishbein and Ajzen's (1975) attitude-behavior model, known as the theory of reasoned action, probably constitutes the most influential and well-documented model of attitudes and decision making (see above). The concepts used in the theory of reasoned action and its successor, the theory of planned behavior (Ajzen, 1991), are similar to the concepts used in another influential theory of human behavior presented above, that is, social cognitive theory (Bandura, 1986).

According to Stone (1998), social cognitive theory stems from social learning theory, which has a rich historical background dating back to the late 1800s, with its early foundation being laid by behavioral and social psychologists and evolved under the umbrella of behaviorism. While there are several versions of social learning theory to which researchers currently subscribe (Stone, 1998), they all share three basic tenets: (a) response consequences (such as rewards or punishments) influence the likelihood that a person will perform a particular behavior again in a given situation (enactive learning); (b) humans can learn by observing others (vicarious learning) in addition to learning by participating in an act personally; and (c) individuals are most likely to model behavior observed by others they identify with.

According to the theory of planned behavior (Ajzen, 1991), human behavior is guided by three considerations: beliefs about the likely consequences or other attributes of the behavior (producing a favorable or unfavorable attitude toward the behavior); beliefs about the normative expectations of other people (resulting in perceived social pressure, i.e., subjective norm); and beliefs about the presence of factors that may further or hinder performance of the behavior (that give rise to perceived behavioral control, the perceived ease or difficulty of performing the behavior).

In origin, the expectancy-value theory and social cognitive theory were focused on different psychological phenomena. Where social learning originally was concerned with learning by observation and imitation (e.g., Miller & Dollard, 1941), the perspective gradually moved from social learning and personality development (e.g., Bandura & Walters, 1963) to behavior modification (e.g., Bandura, 1969) and later on to behavioral change (e.g., Bandura, 1977) with the introduction of the concept self-efficacy, which eventually resulted in Bandura's (1986) social cognitive theory. The expectancy-value theory was originally concerned with the internal processes of human behavior, such as beliefs and attitudes (e.g., Fishbein, 1968; Fishbein & Raven, 1962), which eventually resulted in the theory of reasoned action (Fishbein & Ajzen, 1975), and, by adding the concept of behavioral control, was extended to the theory of planned behavior (Ajzen, 1991).

Conceptually, the concept of perceived behavioral control is by no means new or original to the theory of planned behavior. A similar idea appears in the model of interpersonal behavior (Triandis, 1980), where it takes the form of facilitating conditions. According to Ajzen, however, perceived behavioral control owes its greatest debt to Bandura's work on self-efficacy (Bandura,

1977, 1986, 1997). Perceived self-efficacy refers to "people's beliefs about their capabilities to exercise control over their own level of functioning and over events that affect their lives" (Bandura, 1991, p. 257).

In retrospect, according to Ajzen (2002), the decision to use the term "perceived behavioral control" to denote this component in the theory of planned behavior may have been misleading and should be read as "perceived control over performance of a behavior." The term has sometimes been taken to refer to the belief that performance of a behavior affords control over attainment of an outcome. This is, according to Ajzen, not the intended meaning. Perceived behavioral control simply denotes subjective degree of control over performance of the behavior itself. The distinction here is, according to Ajzen, the same as that between self-efficacy expectations and outcome expectations in social cognitive theory. According to Ajzen, it can be seen that perceived behavioral control and self-efficacy are quite similar: both are concerned with the perceived ability to perform a behavior or sequence of behaviors.

Also, the concept of attitude towards a behavior as used in the theory of reasoned action and theory of planned behavior, which is related to the subjective values of the behavior's perceived outcomes—that is, outcome expectations (Ajzen, 2002)—is similar to outcome expectations defined as judgments of the likely consequence of behavior (Bandura, 1997) within social cognitive theory.

As described earlier in this chapter, the unified theory of acceptance and use of technology is constructed out of eight models. However, the theoretical background of these models originates also from either social cognitive theory, the theory of reasoned action, the theory of planned behavior, or a combination of these theories. Even the diffusion of innovations theory is related to social cognitive theory; see, for example, Bandura's (2006) book chapter "On Integrating Social Cognitive and Social Diffusion Theories."

The above-mentioned theoretical connections and similarities between the three theoretical perspectives on communication technology adoption behavior clearly indicate the convergence between the theoretical perspectives with regard to the theoretical concepts used in the three communication technology adoption models. As a result of the theoretical connections and similarities between the three theoretical perspectives with regard to the central processes and phenomena of interest, it would be interesting to compare the differences in performance of the three communication technology adoption models to explain and predict communication technology adoption behavior.

A Comparison of the Three Models to Explain and Predict Communication Technology Adoption

An examination of the background theories from which the three communication technology adoption models originate indicate there are several theoretical connections and similarities between the three theoretical perspectives with regard to the central processes and phenomena of interest. Yet, all three models propose different determinants to explain and predict communication technology adoption behavior.

Briefly stated, from a uses and gratifications research perspective, communication technology adoption behavior is determined by needs that generate expectations of the media technology, which lead to differential patterns of media technology exposure, resulting in need gratifications and other consequences (cf. Katz, Blumler & Gurevitch, 1974). Communication technology adoption behavior from an expectancy-value perspective that has been incorporated into uses and gratifications research is determined by the behavioral intention to adopt communication technology. Behavioral intention is, in turn, a function of attitude towards the behavior (the sum of the perceived values of the need-driven expectations of the media technology), subjective norms, and, in case the attainment of behavioral goals, is not under volitional control, perceived behavioral control (cf. Ajzen, 1991; Fishbein & Ajzen, 1975). From a social cognitive perspective, communication technology adoption behavior is determined by exercising forethought (e.g., outcome expectations), reflecting on one's own behavior (e.g., self-efficacy), applying self-reactive motivating influences (e.g., self-regulation, habit), and the interaction with the environment (cf. Bandura, 2002). Within the unified theory of acceptance and use of technology, communication technology adoption behavior is determined by performance expectancy, effort expectancy, social influence, and facilitating conditions (Venkatesh et al, 2003).

Peters (2007) argues that it makes quite a difference which determinants are most important to explain communication technology adoption behavior because the explanatory significance of a particular determinant all depends on the stage of development and diffusion the communication technology is in (e.g., a well-accepted media technology, such as the mobile phone, versus a new innovative media technology, such as mobile video phone). Peters both empirically and theoretically compared the three above-mentioned models within the context of mobile communication technology adoption. The findings of Peters' study indicate that each of the three models has its own quality in explaining mobile communication technology adoption behavior. Each model has its own unique contribution in either the explanation or prediction of communication technology adoption behavior.

Expectancy-Value Judgments Model of Uses and Gratifications

The expectancy-value judgments model of uses and gratifications is most useful to explain communication technology adoption behavior in terms of expectations and beliefs as it describes what kind of gratifications people seek (reformulated as expectancy-value judgments) in using a particular media technology. However, for each new media technology that is introduced one needs to establish first and foremost the expectancy-value judgments clusters that match the use of this particular new media technology. Therefore, the strength of the expectancy-value judgments model of uses and gratifications is its capacity for comparative analysis to explain differences in expectations and beliefs among different already accepted communication technologies (e.g., mobile phone use versus landline phone use) for different audiences (cf. McQuail, 2001).

Malhotra (2005) posed that the variables that moderate the effect of attitudes and intentions need to be identified and understood. Moderators that have been identified include motivation and ability, experience, prior knowledge, and mere exposure (Haugtvedt, 1997). According to Malhotra (2005), theories such as the theory of reasoned action and the theory of planned behavior (the background theories of the expectancy-value judgments model of uses and gratifications) need to be modified to account for automatic habitual behavior, such as that engendered by past use.

The usefulness of the expectancy-value model of uses and gratifications to predict the adoption of a new communication technology is not sufficient. Because not all corresponding expectancy-value judgments clusters that might be important to a new communication technology are known beforehand, one must solely rely on previously established gratification clusters of other (perhaps related) communication technologies. Furthermore, Peters (2007) argues that within the expectancy-value model of uses and gratifications a person's attitude towards the intention to use a new communication technology has no significant influence on intention. Peters' findings support the proposition that attitudes do not necessarily lead to intentions to act because they frequently fail to contain sufficient motivational content. That is, evaluations of the consequences of acting or of acting, per se, express one's liking towards a behavior but do not necessarily imply a motivational commitment to act (Bagozzi, 1992; Bagozzi & Kimmel, 1995).

Unified Model of Acceptance and Use of Technology

The unified model of acceptance and use of technology is most useful to supply general information on users' opinions about a technological innovation already present in the organization in terms of expectancies and conditions of the adoption. Peters (2007) argues that although statistically the unified model of acceptance and use of technology surpasses the expectancy-value model of uses and gratifications and the model of media attendance in explaining and predicting mobile communication technology behavior, the specific quality of the constructs of the unified model limits its generalizability beyond an organizational context. Therefore, employing the specific constructs of the unified model to explain and predict general communication technology adoption behavior might not be theoretically justifiable, unless the measures of the unified model are revalidated in the new context one wishes to use the unified model in.

According to Venkatesh et al. (2003), the measures for the unified theory of acceptance and use of technology should be viewed as preliminary, and future research should be targeted at more fully developing and validating appropriate scales for each of the constructs with an emphasis on content validity and then revalidating the model specified herein (or extending it accordingly) with new measures (p. 468). Peters (2007) argues that the explanatory power of the unified model can be improved by incorporating theoretical meaningful explanatory variables that could further explain the dynamics between the different constructs of the unified model of acceptance and use of technology. Parsimony—the need to identify factors that account for the most variance—is to be greatly valued (Burgoon & Buller, 1996); however, not at the expense of explanatory power. Therefore, the unified model as it is originally defined by Venkatesh et al. is more useful to predict users' general opinions about expected use (e.g., opinions about the expected performance and required effort to use a particular technology) than to explain the motivations related to the continued and increased adoption of a particular communication technology. This is because the unified model cannot explain the different underlying mechanisms such as how people come to learn of and choose to initially adopt a technological innovation (cf. Stafford et al., 2004, p. 265).

Model of Media Attendance

Although, the model of media attendance was originally proposed to explain the determinants of media technology usage in social cognitive terms, Peters (2007) argues that the model of media attendance is also capable to explain the determinants of people's intention to adopt a new media technology. In terms of explaining and predicting communication technology adoption behavior, Peters showed that, within the model of media attendance, existing communication technology use (e.g., using a mobile phone) is more likely to be explained by habit strength, and the intention to adopt new communication technology (e.g., the intention to use a mobile video phone) is more likely to be predicted by outcome expectations. This finding implies a reciprocal relationship between habit strength and expected outcomes. Once communication technology adoption is more strongly determined by habituation, the effect of outcome expectations in determining people's communication technology adoption behavior may no longer have much influence because people are no longer aware of the relative importance of expected outcomes or no longer have expectations because the outcomes are already known. On the other hand, the results of Peters' study showed a strong effect of expected outcomes on prospective habit strength with regard to predicting the adoption of a new communication technology. This finding implies that as long as communication technology adoption is not fully habitualized, habit strength is causally determined by outcome expectations, which precede habit strength in time (cf. LaRose & Eastin, 2004). Future research should investigate more in-depth this reciprocal relationship, which requires a longitudinal research design.

The model of media attendance grounded in social cognitive theory offers some promising steps forward in the understanding of the underlying mechanisms that determine communication technology adoption behavior from the perspective of explaining and predicting communication adoption behavior as well as from the perspective of validating and extending theory about the adoption and use of communication technology. More stringent follow-up tests of the model of media attendance extended to other communication technologies and within different contexts of communication adoption are needed to further state the degree of corroboration of the model of media attendance.

Conclusions

This chapter focused on three models to explain and predict communication technology adoption behavior from different theoretical perspectives. The most important implication that can be derived from this chapter is that the choice of a model to understand a particular communication technology adoption behavior should be determined foremost by the stage of development and diffusion the particular communication technology is in. Depending on whether the communication technology in question is an already well-accepted communication technology or a new innovative communication technology, an appropriate communication technology adoption model should be selected to respectively explain and predict communication technology adoption behavior.

Although the discussed communication technology adoption models in this chapter have all been used to explain and predict communication technology adoption behavior, only the model of media attendance has proven to be capable to theoretically explain the dynamics of communication technology adoption behavior in the context to explain existing communication technology use and to predict the intention to adopt a new communication technology (Peters, 2007). In contrast with the two alternative models to understand communication technology adoption behavior, the theoretical core of the model of media attendance remains the same for all communication technologies in explaining and predicting communication technology adoption behavior. As the components of the model of media attendance are constructed to represent general mechanisms that determine human behavior, only the underlying indicators of the construct have to be reformulated in the context of the particular communication technology of interest. Whereas in the expectancy-value judgments model of uses and gratifications for each particular communication technology, first a tailor-made list of expectancy-value judgments clusters have to be developed and validated before this particular communication technology can be examined by the model. Likewise, because of the limitation in generalizability beyond its original context, the measures of the unified model of acceptance and use of technology also have to be revalidated first before the unified model can be used to examine a particular communication technology that lies outside of the model's original scope.

In sum, it can be concluded that (a) the model of media attendance should perform well in explaining and predicting communication technology adoption behavior due to the model's underlying interrelated and reciprocal mechanisms of general human behavior; (b) the expectancy-value judgments model of uses and gratifications should perform well in explaining existing communication technology behavior in terms of beliefs and expectations; and (c) the unified model of acceptance and use of technology should perform well in supplying general information on users' opinions about a technological innovation in terms of expectancies and conditions of use.

References

Aarts, H., Verplanken, B., & Van Knippenberg, A. (1998). Predicting behaviour from actions in the past: Repeated decision making or a matter of habit? *Journal of Applied Social Psychology* 28(15), 1355–1374.

Ajzen, I. (1991). The theory of planned behavior. *Organizational Behavior and Human Decision Processes*, 50, 179–211.

Ajzen, I. (2002). Perceived behavioral control, self-efficacy, locus of control, and the theory of planned behavior. *Journal of Applied Social Psychology*, 32(4), 665–683.

Arkesteijn, K., & Oerlemans, L. (2005). The early adoption of green power by Dutch households: An empirical exploration of factors influencing the early adoption of green electricity for domestic purposes. *Energy Policy*, 33(2), 183–196.

Babrow, A. S. (1989). An expectancy-value analysis of student soap opera audience. *Communication Research*, 16, 155–178.

Babrow, A. S., & Swanson, D. L. (1988). Disentangling antecedents of audience exposure levels: Extending expectancy-value analyses of gratifications sought from television news. *Communication Monographs*, 55, 1–21.

Bagozzi, R. P. (1992). The self-regulation of attitudes, intentions and behavior. *Social Psychology Quarterly*, 55, 178–204.

Bagozzi, R. P., Davis, F. D., & Warshaw, P. R. (1992). Development and test of a theory of technological learning and usage. *Human Relations*, 45(7), 660–686.

Bagozzi, R. P., & Kimmel, S. K. (1995). A comparison of leading theories for the prediction of goal-directed behaviors. *British Journal of Social Psychology*, 34, 437–461.

Bandura, A. (1969). *Principles of behavior modification*. New York: Holt, Rinehart, & Winston.

Bandura, A. (1977). *Social learning theory*. Englewood Cliffs, NJ: Prentice Hall.

Bandura, A. (1986). *Social foundations of thought and action: A social cognitive theory*. Englewood Cliffs, NJ: Prentice Hall.

Bandura, A. (1991). Social cognitive theory of self-regulation. *Organizational Behavior and Human Decision Processes*, 50, 248–287.

Bandura, A. (1997). *Self-efficacy: The exercise of control.* New York: Freeman.

Bandura, A. (2002). Social cognitive theory of mass communication. In J. Bryant and D. Zillman (Eds.), *Media effects: Advances in theory and research* (2nd ed., pp. 121–155). Mahwah, NJ: Lawrence Erlbaum.

Bandura, A. (2006). On integrating social cognitive and social diffusion theories. In A. Singhal & J. W. Dearing (Eds.). *Communications of innovations: A journey with Ev Rogers* (pp. 111–135). New Delhi, India: Sage.

Bandura, A., & Walters, R. H. (1963). *Social learning and personality development.* New York: Holt, Rinehart & Winston.

Book, C. L., & Barnett, B. (2006). PCTV: Consumers, expectancy-value and likely adoption. *Convergence: The International Journal of Research into New Media Technologies 12*(3), 325–339.

Bryant, J., & Miron, D. (2004). Theory and research in mass communication. *Journal of Communication, 54*(4), 662–704.

Bryant, J., & Zillman, D. (Eds.). (2001). *Media effects: Advances in theory and research* (2nd ed.). Hillsdale, NJ: Lawrence Erlbaum.

Burgoon, J. K., & Buller, D. B. (1996). Reflections on the nature of theory building and the theoretical status of interpersonal deception theory. *Communication Theory*, 6, 311–328.

Compeau, D. R., & Higgens, C. A. (1995). Computer self-efficacy: Development of a measure and initial test. *MIS Quarterly, 19*(2), 189–211.

Cronkhite, G. (1969). *Persuasion: Speech and behavioral change.* Indianapolis: Bobbs-Merrill.

Dabholkar, P. A., & Bagozzi, R. P. (2002). An attitudinal model of technology-based self service: Moderating effects of consumer traits and situational factors. *Journal of the Academy of Marketing Science*, 30(3), 184–201.

Davis, F. D. (1989). Perceived usefulness, perceived ease of use, and user acceptance of information technology. *MIS Quarterly, 13* (3), 319–340.

Davis, F. D., Bagozzi, R. P., & Warshaw, P. R. (1989). User acceptance of computer technology: A comparison of two theoretical models. *Management Science*, 35, 982–1003.

Eastin, M. S., & LaRose, R. (2000). Internet self-efficacy and the psychology of the digital divide. *Journal of Computer Mediated Communication*, 6(1). Retrieved October 9, 2006, from http://jcmc.indiana.edu/vol6/issue1/ eastin.html

Fichman, R. G., & Kemerer, C. F. (1997). The assimilation of software process innovations: An organizational learning perspective. *Management Science*, 43(10), 1345–1363.

Fishbein, M. (1968). An investigation of relationships between beliefs about an object and the attitude towards that object. *Human Relationships, 16*, 233–240.

Fishbein, M., & Ajzen, I. (1975). *Beliefs, attitude, intention and behavior: An introduction to theory and research.* Reading, MA: Addison-Wesley.

Fishbein, M., & Raven, B. H. (1962). The AB scales: An operational definition of belief and attitude. *Human Relations, 12*, 32–44.

Flett, R., Alpass, F., Humphries, S., Massey, C., Morris, S., & Long, N. (2004). The technology acceptance model and use of technology in New Zealand dairy farming. *Agricultural Systems*, 80(2), 199–211.

Galloway, J. J., & Meek, F. L. (1981). Audience uses and gratifications: An expectancy model. *Communication Research*, 8, 435–450.

Gentry, L., & Calantone, R. (2002). A comparison of three models to explain shop-bot use on the web. *Psychology and Marketing, 19*(11), 945–956.

Goodhue, D. L. (1995). Understanding user evaluations of information systems. *Management Science, 41*(12), 1827–1844.

Greenberg, B. S. (1974). Gratifications of television viewing and their correlations for British children. In J. G. Blumler & E. Katz (Eds.), *The uses of mass communication: Current perspectives on gratifications research* (pp. 71–92). Beverly Hills, CA: Sage.

Haugtvedt, C. (1997). Beyond fact or artifact: An assessment of Fishbein and Meddlestadt's perspectives on attitude change processes. *Journal of Consumer Psychology,* 6(1), 99–106.

Hendriks Vettehen, P. G. (1998). Conceptualisering en operationalisering van het begrip "motief" in Uses & Gratifications onderzoek. [Conceptualization and measurement of the motivation concept in Uses & Gratifications research]. Doctoral dissertation, University of Nijmegen, Nijmegen: ITS.

Hendriks Vettehen, P. G., & Van Snippenburg, L. B. (2002). Measuring motivations for media exposure: A thesis. *Quality and Quantity,* 36(3), 259–276.

Hofstetter, C. R., Zuniga, S., & Dozier, D. M. (2001). Media self-efficacy: Validation of a new concept. *Mass Communication and Society,* 4(1), 61–78.

Infante, D. A., Rancer, A. S., & Womack D. F. (1997). *Building communication theory* (3rd ed.). Prospect Heights, IL: Waveland.

Katz, E., Blumler, J. G., & Gurevitch, M. (1974). Utilization of mass communication by the individual. In J. G. Blumler & E. Katz (Eds.) *The uses of mass communication* (pp. 19–34). Beverly Hills, CA: Sage.

Katz, E., Gurevitch, M., & Haas, H. (1973). On the use of the mass media for important things. *American Sociological Review,* 38 (April), 164–181.

LaRose, R., & Eastin, M. S. (2004). A social cognitive theory of Internet uses and gratifications: Toward a new model of media attendance. *Journal of Broadcasting and Electronic Media,* 48(3), 358–377.

LaRose, R., Lai, Y. J., Lange, R., Love, B., & Wu, Y. (2005). Sharing or piracy? An exploration of downloading behavior. *Journal of Computer Mediated Communication, 11*(1), Article 1. Retrieved November 21, 2005, from http://jcmc.indiana.edu/vol11/issue1/larose.html

LaRose, R., Lin, C. A., & Eastin, M. S. (2003). Unregulated Internet usage: Addiction, habit, or deficient self-regulation? *Media Psychology,* 5, 224–253.

LaRose, R., Mastro, D. A., & Eastin, M. S. (2001). Understanding Internet usage: A social cognitive approach to uses and gratifications. *Social Science Computer Review, 19*, 395–413.

Lazarsfeld, P., & Stanton, F. N. (1944). *Radio Research, 1942–43*. New York: Duell, Sloan and Pearce.

Lazarsfeld, P. F., Berelson, B. & Gaudet, H. (1948). *The people's choice* (2nd ed.). New York: Columbia University Press.

Leonard-Barton, D., & Deschamps, I. (1988). Managerial influence in the implementation of new technology. *Management Science,* 34(10), 1252–1265.

Leung, L., & Wei, R. (1999). Seeking news from the pager: A value-expectancy study. *Journal of Broadcasting & Electronic Media,* 43(3), 299–315.

Malhotra, N. K. (2005). Attitude and affect: New frontiers of research in the 21st century. *Journal of Business Research,* 58, 477–482.

McQuail, D. (2001). With more hindsight: conceptual problems and some ways forward for media use research. *Communications: The European Journal of Communication Research, 26*(4), 337–350.

Miller, N. E., & Dollard, J. (1941). *Social learning and imitation.* New Haven, CT: Yale University Press.

Palmgreen, P. (1984). Uses and gratifications: A theoretical perspective. In R. N. Bostrom (Ed.), *Communication yearbook* 8 (pp. 20–55). Beverly Hills, CA: Sage.

Palmgreen, P., & Rayburn, J. D. (1985). An expectancy-value approach to media gratifications. In K. E. Rosengren, L. A. Wenner, & P. Palmgreen (eds.), *Media gratifications research* (pp. 61–72). London: Sage.

Palmgreen, P., Wenner, L. A., & Rayburn, J. D. (1980). Relations between gratifications sought and obtained: A study of television news. *Communication Research, 7*, 161–192.

Palmgreen, P., Wenner, L., & Rosegren, K. (1985). Uses and gratifications research: The past ten years. In K. Rosegren, L. Wenner & P. Palmgreen (Eds.) *Media gratifications research* (pp. 11–37). Beverly Hills, CA: Sage.

Peters, O. (2007). *Social psychological determinants of mobile communication technology use and adoption. A comparison of three models to explain and predict mobile communication technology behavior.* Enschede, NL: University of Twente.

Rayburn, J. D., & Palmgreen, P. (1984). Merging uses and gratifications and expectancy-value theory. *Communication Research, 11*, 537–562.

Rogers, E. M. (2003). *Diffusion of innovations* (5th ed.). New York: Free Press.

Rosegren, K. E. (1974). Uses and gratifications: A paradigm outlined. In J. G. Blumler & E. Katz (eds.). *The uses of mass communication* (pp. 269–286). Beverly Hills, CA: Sage.

Rosenberg, M. J. (1956). Cognitive structure and attitudinal effects. *Journal of Abnormal and Social Psychology, 53*, 367–372.

Ruggiero, T. E. (2000). Uses and gratifications theory in the 21st century. *Mass Communication & Society, 3*(1), 3–37.

Sambamurthy, V., & Chin, W. (1994). The effects of group attitudes towards GDSS designs on the decision-making performance of computer-supported groups. *Decision Sciences, 25*(2), 215–242.

Schramm, W. (1954). How communication works. In W. Schramm (Ed.), *The process and effects of mass communication* (pp. 3–26). Urbana, IL: University of Illinois Press.

Stafford, T. F., Stafford, M. R., & Schkade, L. L. (2004). Determining uses and gratifications for the Internet. *Decision Sciences, 35*(2), 259–288.

Stone, D. (1998). *Social cognitive theory.* Retrieved March 19, 2004, from http://hsc.usf.edu/~kmbrown/Social_Cognitive_Theory_Overview.htm

Taylor, S., & Todd, P. A. (1995). Understanding information technology use: A test of competing models. *Information Systems Research*, 6, 144–176.

Thompson, R. L., Higgins, C. A., & Howell, J. M. (1991). Personal computing: Toward a conceptual model of utilization. *MIS Quarterly, 15*(1), 124–143.

Triandis, H. C. (1980). Values, attitudes, and interpersonal behavior. In H. Howe & M. Page (Eds.), *Nebraska symposium on motivation 1979: Beliefs, attitudes, and values* (pp. 195–259). Lincoln, NE: University of Nebraska Press.

Venkatesh, C. (2000). Determinants of perceived ease of use: Integrating perceived behavioral control, computer anxiety and enjoyment into the technology acceptance model.

Information Systems Research, 11(4), 342–365.
Venkatesh, V. (2006). Where to go from here? Thoughts on future directions for research on individual-level technology adoption with a focus on decision making. *Decision Sciences*, *37*(4), 497–518.
Venkatesh, V., & Davis, F. D. (2000). A theoretical extension of the technology acceptance model: Four longitudinal studies. *Management Science*, *46*(2), 186–204.
Venkatesh, V., Morris, M. G., Davis, G. B., & Davis, F. D. (2003). User acceptance of information technology: Toward a unified view. *MIS Quarterly*, *27*(3), 425–478.
Yang, Z. L., & Peterson, R. T. (2004). Customer perceived value, satisfaction, and loyalty: The role of switching costs. *Psychology and Marketing*, *21*, 799–822.

· 3 ·

Overcoming the Binary Logic of Adoption

On the Integration of Diffusion of Innovations Theory and the Concept of Appropriation

VERONIKA KARNOWSKI, THILO VON PAPE & WERNER WIRTH

Introduction

When opting for communication studies as a basis to integrate existing traditions of diffusion research, Rogers (1962) had fed the complex field of research with a rich set of theoretical and methodological approaches. Today, diffusion research is struggling again to meet the complexity of innovations and their uses—namely in the field of new information and communication technologies. Hence, a realignment to recent progresses of communication studies and neighboring disciplines is recommended.

The chapter demonstrates how the traditionally one-directional and binary concepts of diffusion and adoption can be complemented by a closer look at implementation as a process of personal and social appropriation. While Rice and Rogers (1980) concede that users reinvent innovations in the course of implementation, scholars have since neglected this point to the profit of ever more elaborated models for predicting individual adoption decisions (Theory of Planned Behavior, Technology Acceptance Model, etc.) and social processes of diffusion (Social Network Analysis). The chapter discusses how to overcome these drawbacks by combining adoption and diffusion approaches with traditions that permit to consider the implementation of innovations as a dynamic, multi-dimensional process.

> I remember interviewing an Iowa farmer during my Ph.D. dissertation research at Iowa State University in 1954. I inquired about his adoption of 2,4-D weed spray. The farmer described in some detail the particular and unusual ways in which he used the weed spray on his farm: At the end of his lengthy remarks, I simply checked "adopter" in my interview questionnaire. —Everett Rogers (2003, p. 17)

As illustrated by this anecdote from Everett Rogers' early fieldwork, empirical diffusion research has always had a somewhat limited view on adopters and what they do with innovations. The present chapter responds to this shortcoming on several levels: by describing how mainstream diffusion theory, from its earliest days, has considered adoption from a binary perspective; by explaining the methodological and theoretical reasons for this approach; by showing how the phenomenon of Re-Invention has always challenged this view, even more so today with the increasing complexity of innovations; and finally by outlining approaches that may lead to a new and broader perspective on what users do with innovations in the course of the diffusion process.

The Evolution of Diffusion of Innovations Theory and Its Binary Logic

A closer look at the history of diffusion research reveals that the idea of adoption as a binary decision cannot be traced back to the very beginning. The French sociologist Gabriel de Tarde, generally considered a forefather of diffusion research, had a much wider understanding of what constitutes an innovation and what users could do with an innovation. Thinking in terms of metaphors rather than hypotheses and operationalizations, Tarde considers innovations as waves that spread among society. Using this metaphor, he expresses two advanced but long-forgotten ideas:

- Innovations may change in the course of diffusion. This may happen when they reach new users (just as waves change their form when entering a new medium) or when they interfere with other innovations.
- The importance of adopters for the evolution of innovations is much greater than generally thought. In an extreme case, the inventor generates only a quasi-accidental impulse that develops its force in the course of spreading throughout society, just as a butterfly flapping its wings may lead to a landslide, in the words of Tarde (1902, p. 562).

These ideas have been largely forgotten (the article in question has yet to be translated into English), as most scholars focus on Tarde's early reflections on opinion leaders and the S-like shape of the adoption curve. The reason for this limited view on Tarde may lie in the way in which diffusion research has evolved since his time.

The first half of the twentieth century was marked by a variety of parallel traditions in various disciplines ranging from anthropology (Wissler, 1914) and rural sociology (Ryan & Gross, 1943) to public health and medical sociology (Menzel & Katz, 1955; Rogers, 2003, pp. 44–45), involving a wide spectrum of theoretical and methodological approaches. However, when Everett Rogers consolidated these approaches to what is considered today as "the traditional approach" (Dearing & Meyer 2006, p. 39), he opted for a rather tight framework, both theoretically and methodologically.

With respect to methodology, Rogers leaned on the practice adopted in agricultural sociology, which he knew from his own studies in this field at Iowa State University, in particular the seminal study by Iowa State sociologists Bryce Ryan and Neal Gross (Ryan & Gross, 1943; cf. Lowery & DeFleur, 1995; Meyer, 2004; Rogers, 2003). In the summer of 1941, Ryan and Gross interviewed 259 farmers regarding their adoption of a new hybrid seed corn during the 1930s. Drawing on these data, they identified factors that influence a single adoption decision and proposed an S-shaped diffusion curve.

Meyer (2004, p. 59) summarizes the applied methodology in terms of the following five points: "1. quantitative data, 2. concerning a single innovation, 3. collected from adopters, 4. at a single point in time, 5. after widespread diffusion had already taken place." This methodology enables a description of the evolution of cumulated adoption decisions over time, but it requires a reduction of the farmers' activity to a simple dichotomy between adoption and rejection. Nonetheless, the authors were themselves highly skeptical as to whether it would be possible to formulate the general shapes of diffusion curves, as the diffusion processes relevant to various innovations are vastly different in terms of the nature of communication among the adopters: "but it seems doubtful if any theoretic pattern can adequately conform to situations involving all degrees of interaction and isolation; to economic practices as well as to styles" (Ryan & Gross, 1943, p. 24). Although Ryan and Gross applied this caveat mainly to the question of which shape a diffusion curve may take (e.g., normal distribution, logistic curve), it can also be understood as a general warning against reducing the complex evolution of diverse innovations to all-embracing theories, not to mention mathematical models.

While Ryan and Gross's approach appears satisfactory in terms of the specific question regarding how to spread a singular, unalterable, and clearly superior innovation among a hesitant population, it neglects many aspects that become more important for other innovations: How do users implement the innovation into their everyday practices? What symbolic meaning does the innovation have for them? How do they communicate these aspects among each other (and not simply communicate the adoption decision)? When Everett Rogers chose to adopt the Ryan and Gross methodology as a basis for consolidating the various existing streams of diffusion research to a single integrative approach, he initiated a development that led to a long-lasting neglect of these questions and cemented the focus on binary adoption decisions (be it on the level of individual adopters or on the cumulative level of social systems) (Meyer, 2004).

This narrowing of the perspective was mirrored by the theoretical basis on which Rogers built the new diffusion research. In the course of looking over various disciplines for a theoretical harbor for his new Diffusion of Innovations Theory, Rogers opted for the relatively young field of communication studies. As noted by Dearing and Singhal (2006, p. 20), the lack of theoretical complexity was the main advantage of Rogers' decision, as communication studies was "sufficiently new and undetermined" to permit a new research field to evolve. The main theoretical achievement of communication that Rogers incorporated into Diffusion of Innovations Theory was the well-known Lasswell formula (Lasswell, 1948).

In their book with the programmatic title *Communication of Innovations*, Rogers and Shoemaker (1972, p. 20) propose a diffusion process parallel to the SMCRE model (Source–Message–Channel–Receiver–Effect): the inventor replaces the "source," the innovation the "message," diffusion channels the "channels," the adopter the "receiver," and adoption the "effects." While this formula fully satisfied the need for a loose structure capable of integrating a large variety of results, this approach also cemented a strongly linear understanding of communication and, hence, of the communication of innovations.

Communication studies have since come up with new models that attribute a more active role to the receiver/adopter and that consider the message as the object of negotiation between the sender and receiver (Hall, 1980). However, at its heart, diffusion research has kept true to the linear Lasswell formula, with a linear understanding of diffusion from the inventor to the late adopters, with a static understanding of the innovation and with a passive perspective on the user, whose only choice of action lies between adoption and rejection.

This approach has the advantage of opening diffusion research to other approaches that work based on binary variables, such as Social Network Analysis or the Theory of Planned Behavior. For example, Social Network Analysis (SNA, Coleman, Katz & Menzel, 1957; Valente, 2006) helps to trace the channels through which innovations spread within interpersonal networks. Behavioral theories from social psychology, such as the Theory of Reasoned Action (TRA, cf. Fishbein & Ajzen, 1975) and the Theory of Planned Behavior (TPB, cf. Ajzen, 1985), facilitate the modeling of factors that influence the individual's adoption decision.

Despite these theoretical advancements, Katz (1999, p. 145) states "There is an apparent paradox at work: the number of diffusion studies continues at a high rate while the growth of appropriate theory is at an apparent standstill."

Re-Invention Challenging Diffusion of Innovations Theory

In the 1970s the adoption process was differentiated by the concept of "Re-Invention," a term used mainly in political science (see Charters & Pellegrin, 1973; Glick & Hays, 1991; Goodman & Steckler, 1989; Hays, 1996a, 1996b; Lewis & Seibold, 1996) and seldom applied in communication science (Rice & Rogers, 1980; Rogers, 2003; Schenk, Dahm & Sonje, 1996). Re-Invention extends the binary adoption decision to a possible change in the innovation during its implementation and analyzes the factors that support a high degree of modification (see Glick & Hays, 1991; Hays, 1996a, 1996b; Lewis & Seibold, 1996; Majchrzak, Rice, Malhotra, King & Ba, 2000; Orlikowski, 1993; Rice & Rogers, 1980; Rogers, 2003).

However, the accomplishments of Diffusion of Innovations Theory in analyzing Re-Invention remain extremely limited. Rogers (Rice & Rogers, 1980; Rogers, 2003, pp. 180–188) focuses mainly on identifying those factors that influence whether Re-Invention will occur and assessing its consequences. On the innovation side, complexity and a loosely bundled structure, as well as a wide spectrum of potential uses, make Re-Invention more likely. On the users' side, either ignorance or a very high technical knowledge (von Hippel, 1986) are often accompanied by Re-Invention, as well as local pride of ownership (Havelock, 1974) or a complex inner structure of the adopting unit (such as a large company adopting a new communication system and needing to adapt the system to its own hierarchy and work processes). The third instance that could push Re-Invention is a change agent, such as a company that counts on its users to actively integrate the products into their everyday lives. This idea is expressed by the use of a slogan such as "You make it a Sony." Finally, Re-Invention also depends on the amount of time that has passed in the diffusion process: Rogers considers that Re-Invention is more likely to occur late in the diffusion process, when members of the social system have become accustomed to the innovation.

The evaluation of Re-Invention is double-edged: on the one hand, it could represent a danger for the user (i.e., when security-sensitive innovations such as electronic devices are re-invented), and it could threaten the producer's business plans. This is the case for the Re-Invention of the "missed call" function for mobile telephones, which is used to communicate without paying for a phone connection (Donner, 2007). On the other hand, Rice and Rogers (1980, p. 504) emphasize the positive effects of Re-Invention when it serves for adapting an innovation to a user's specific needs, leading to a higher likelihood of confirmation after the implementation phase. Hence, Rogers acknowledges the existence of Re-Invention and explores its causes and consequences. However, this acknowledgment requires a rethink of the linear concept of diffusion and its replacement with a communication model that assigns an active role to the recipient/adopter, although this has yet to be done in diffusion research. Instead, the findings about Re-Invention only serve as a marker that highlights the boundaries of traditional diffusion research. Therefore, we need to look beyond diffusion research to find ways of overcoming existing limitations to diffusion of innovations theory.

Recent digital, interactive, and/or mobile media innovations commonly constitute a bundle of technological functions and services that lead to a variety of applications. Such technologies are commonly status symbols of high relevance to the user's self-perception (see Ling, 2004; Oksman & Turtiainen, 2004). Consequently, these technologies, from early word processing software (Johnson & Rice, 1984) to today's mobile telephones (Donner, 2007), are likely to be re-invented by users. Until now, diffusion of innovations theory has been unable to describe this change beyond the concept of Re-Invention, largely concentrating on the factors that support modification.

Overcoming the Binary Logic of Adoption

In response to the phenomenon of Re-Invention, Rogers (2003, p. 181) suggests that "The fact that Re-Invention may happen is a strong argument for measuring adoption at the implementation stage, as change that has really happened." Because implementation is a more complex phenomenon than adoption, this demand would consequently mean that we must go beyond the binary logic of adoption. Accordingly, research into the "change that has really happened" would be more than just stating whether a certain user adopted an innovation: it should be about analyzing the circumstances and effects of the users' creativeness in the course of making sense of the innovation in question, i.e., appropriating the innovation. Below, we consider theories and findings from other research traditions that are highly compatible with Diffusion of Innovations Theory and that may aid our understanding of "adoption at the implementation stage, as change that has really happened" (Rogers, 2003, p. 181).

Cultural Studies

Cultural Studies emphasizes the importance of interpersonal communication when adopting innovations. In the nomenclature of this field, this is called the process of appropriation. Media appropriation is a constructive process of sense-making among the users of the content of any media, including, for example, text in literature (see Certeau, 2002) or television (Brown, 1994; Hall, 1980). These ideas basically rest upon the Encoding/Decoding Model proposed by Hall (1980). In opposition to any linear communication model, Hall postulates that the transmission of a certain message is a process of encoding and decoding. The sender first encodes the message in a certain set of symbols before the symbols are decoded by the receiver ("appropriated"). Only once the receiver has decoded the transmitted message may any effects of the message occur. Crucial to this idea of encoding and decoding is the assumption that these processes are shaped by the sender's and receiver's respective meaning structures. As these meaning structures are shaped by the social environment and the communication processes, the sender's and receiver's meaning structures may be quite different. Consequently, the meaning of a message could be altered during this process of transmission.

Hall (1980) describes the following three types of alteration ("readings") during the communication process. First, in the *dominant/hegemonic reading*, the sender's and receiver's meaning structures are similar enough that the message's intended meaning is retained, i.e., the receiver decodes the message in the same sense as it was encoded by the sender. When the receiver decodes the message in the opposite way to the sender's intentions, this is called *oppositional reading*. Finally, *negotiated reading* is intermediate between dominant and oppositional reading, whereby the receiver in general decodes the message in the overall intended sense but makes certain alterations.

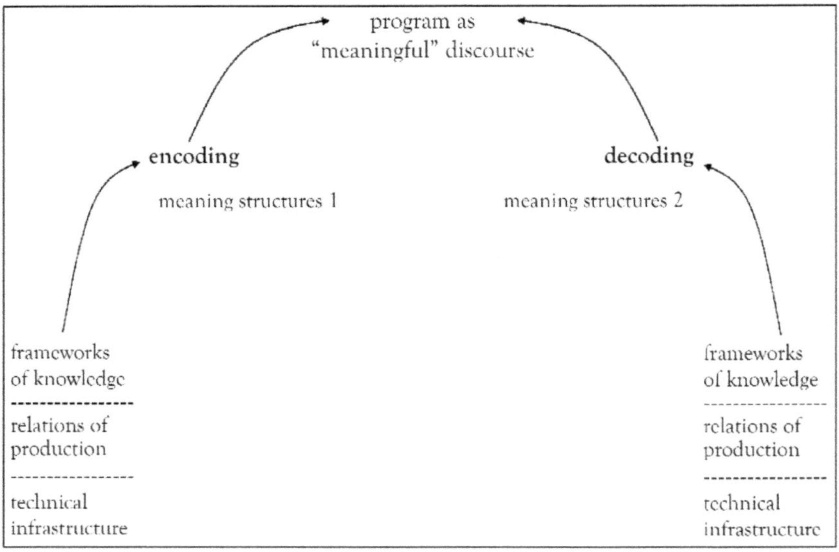

Figure 1: Encoding/Decoding Model (Hall, 2003, p. 168).

Silverstone and Haddon's (1996) Domestication Approach widens this concept of appropriation used in Cultural Studies by adapting it to the adoption of new information and communication technologies in the household (see also Lehtonen, 2003; Oksman & Turtiainen, 2004; Quandt & von Pape, 2010). The authors develop a dynamic concept of appropriation. In a first dimension, called *Commodification*, the potential user gains a first impression of the innovation based on the input from advertising, mass media, and other users. The second dimension, *Appropriation* itself, contains a spatial and a temporal aspect. Objectification is spatial appropriation; thus, the object must be placed in the user's apartment. Incorporation, on the other hand, deals with the temporal integration of an innovation in the user's habits. The third dimension, *Conversion*, is about the user's self-expression based on the innovation.

As with the concept of Re-Invention, Cultural Studies point out that an innovation doesn't necessarily have to be understood and "appropriated" by the adopter in the way intended by the developer. Rather, each adopter has a unique style of appropriating a certain innovation, depending on the second point brought up Cultural Studies: the negotiation of meaning. According to Hall's (1980) idea regarding "meaning structures" shaping the individual appropriation process, the adopter's communication with his or her social environment (i.e., meta-communication about the innovation) affects the way in which the adopter implements the innovation (see also Frissen, 2000; Ling, Nilsen & Granhaug, 1999; Silverstone & Haddon, 1996).

Social Learning Theory

Research into diffusion at the implementation stage involves analysis of the spread of certain forms of behavior. Social Learning Theory is based on the idea of vicarious learning (i.e., adopting certain forms of behavior) by observing modeled behavior supported by different motivational aspects. Bandura (1977) describes four sub-processes of observational learning (see Figure 1):

1. Attentional processes manage the selection of modeled events. This process is influenced by salience, affective valence, complexity, prevalence, accessibility, and the functional value of the modeled event, as well as by the observer's perceptual set, cognitive capabilities, cognitive preconceptions, arousal level, and acquired preferences (see Bandura, 1977, 1986).

2. The retention of modeled events is the second precondition for observational learning. Thus, the observer must transform the modeled events into cognitive structures by the use of symbols. In this way, the retention is facilitated by repeated observation of a modeled event (see Bandura, 1977, 1986).

3. Production processes are concerned with the transfer from cognitive structures to behavior. In this context, Bandura emphasizes that the individual can rearrange the learned behavioral elements, thereby creating new behavioral patterns:

   > "When exposed to models who differ in their styles of thinking and behavior, observers rarely pattern their behavior exclusively after a single source, nor do they adopt all the attributes even of preferred models. Rather, observers combine various aspects of different models into new amalgams that differ from the individual sources" (Bandura, 1986, p. 104).

4. Motivational processes control which observed events are imitated. These motivational aspects may be external or vicarious incentives. The individual may also reward him/herself (self-incentives) by anticipating positive outcomes arising from the behavior in question (see Bandura, 1977, 1986).

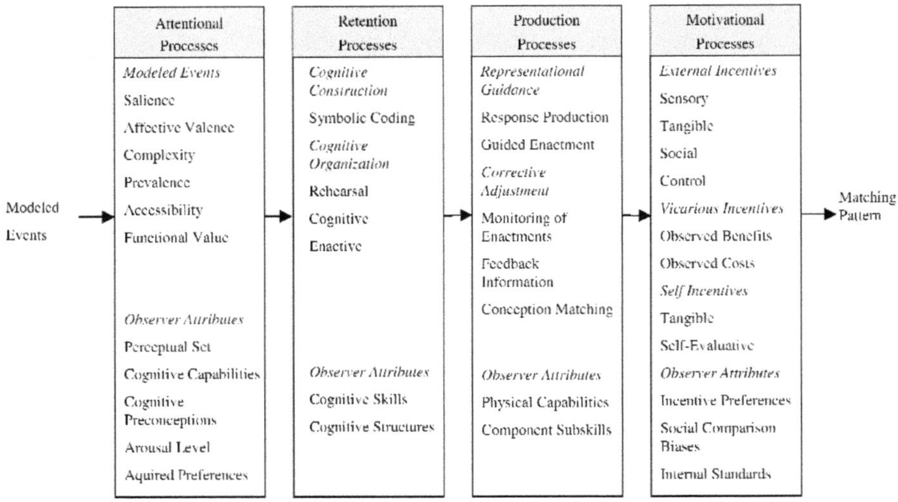

Figure 2: Four Sub-processes of Observational Learning (Bandura, 2001, p. 273).

According to Bandura (1986), the modeled events may be either observed in the direct surroundings of an individual or in the mass media, without any basic difference in the resulting effects. Observation in the mass media is termed "symbolic modeling." Although Bandura (1986) doesn't define the term "media", it can be argued that he acts on the assumption of mass media, as he defines symbolic models by their transmission to a large number of recipients: "it can transmit simultaneously knowledge of wide applicability to vast numbers of people through the medium of symbolic models" (Bandura, 1986, p. 47).

Consequently, Social Learning Theory could add different aspects to Diffusion of Innovations Theory. First, it agrees with Cultural Studies regarding the importance of the cultural environment in the implementation stage. Second, it adds the dimension of mass communication as a source of modeled behavior. Rogers himself argues that imitation of modeled behavior is a central process in diffusion: This interdependence on the experience of near peers suggests that the heart of the diffusion process consists of the modeling and imitation by potential adopters by their network partners who have previously adopted. Diffusion is a very social process that involves interpersonal communication relationships (Rogers, 2003, p. 19).

This implicit connection to Social Learning Theory is picked up by Bandura (2006) himself, who also argues that symbolic modeling is a core process of diffusion (Bandura, 2006, p. 125) and refers to the creative side of the adopter in the diffusion process:

> Creativeness rarely springs entirely from individual inventiveness. Indeed, Selective modeling is often the mother of invention. People adopt the modeled beneficial elements, improve upon them, synthesize them into new forms, and tailor them to their particular circumstances. (Bandura, 2006, p. 117)

Consequently, Bandura uses the term *social diffusion* rather than diffusion. Although these acknowledgments were widely used in establishing edutainment TV and radio serials supporting healthy lifestyles in developing countries (see Rogers, Vaughan, Swalehe, Rao, Svenkerud & Sood, 1999; Singhal, Cody, Rogers & Sabido, 2004; Vaughan & Rogers, 2000), the influence of modeling (interpersonal and via mass media) on the diffusion of media innovations is generally overlooked.

MPA model

The Mobile Phone Appropriation Model (MPA model) (Wirth, von Pape & Karnowski, 2008) deals with the question of how mobile phones are integrated into the user's daily routine. The model was developed on the basis of adoption and diffusion research (e.g., Diffusion of Innovations Theory, Rogers, 2003; Theory of Planned Behavior, Ajzen, 2005; Technology Acceptance Model, Davis, 1986) as well as appropriation research (e.g., Frame Analysis, Goffman, 1974; Domestication Approach, Silverstone & Haddon, 1996; Uses-and-Gratifications Approach, Katz, Blumler & Gurevitch, 1974). It can be understood as an extension of the model applied in Theory of Planned Behavior (Ajzen, 2005), which differentiates the TPB model's structure in order to grasp the various forms of appropriation (see Figure 2; for a detailed description see Wirth, von Pape & Karnowski, 2008):

1. The model considers appropriation to be a creative and active process, ending up in individual usage and meaning patterns. Thus, behavior is differentiated into its object-related and functional aspects. The object-related aspects include fashion aspects (e.g., ring tones and accessories), handling aspects, and the general usage frequency of different functionalities such as telephony, text messaging, and online services. The functional aspects represent the large variety of uses of the mobile telephone known from appropriation research and the Uses-and-Gratifications Approach (e.g., the management of daily life, maintaining relations; see

Höflich & Rössler, 2001; Leung & Wei, 2000), with an emphasis on the symbolic dimension (e.g., status; see Leung & Wei, 2000) of this process.

2. The model takes into account the symbolic value of the object mobile telephone and its usage.
3. The model no longer takes TPB's independent variables of behavioral beliefs, normative beliefs, and control beliefs to be static; instead, they are understood as the constantly evolving results of the appropriation process (Jonas & Doll, 1996; Kendzierski, 1990). Consequently, the model is conceptualized as a cycle, with appropriation being a constantly renewed process. Pragmatic and symbolic use is not only the result of behavioral, normative, and control beliefs, but also their basis (Ouellette & Wood, 1998).
4. The model comprises the impact of communication on the appropriation process: meta-communication. Thus, behavioral, normative, and control beliefs, as well as symbolical and practical behavior, are negotiated via communication among users, producers, and mass media, be it mass communication, personal influences, or the simple demonstration and observation of one's mobile phone use.

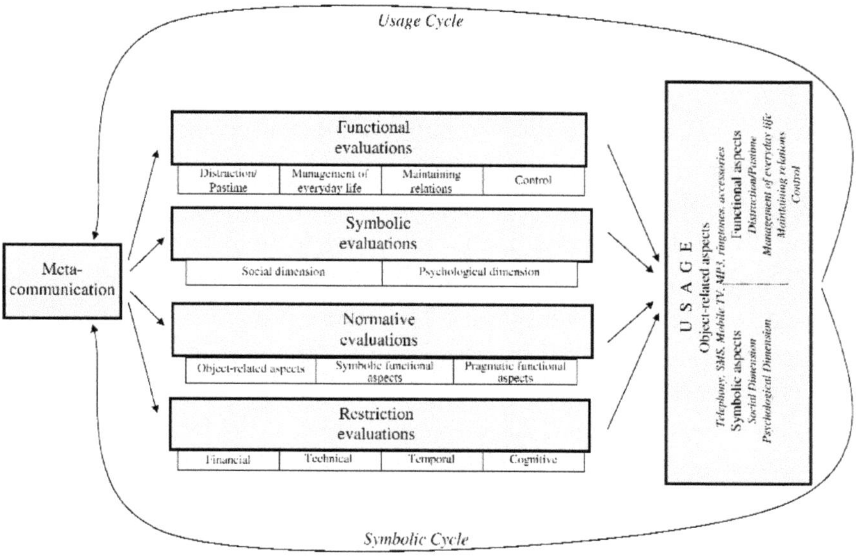

Figure 3: The MPA Model (Wirth et al., 2008, p. 606).

Similar to Cultural Studies and Social Learning Theory, the MPA model highlights the importance of taking into account both meta-communication and the user's creative input in the diffusion process. Although the MPA model was developed and empirically tested (von Pape, Karnowski & Wirth, 2008) to explain the appropriation of mobile phones, it has also proved successful in explaining the appropriation of other recent media innovations (Karnowski & von Pape, 2009; von Pape, Karnowski, Wirth, Klimmt & Hartmann, 2007).

Conclusion

There exists ample empirical evidence of the changing of innovations during the diffusion process; indeed, such change appears to be becoming an increasingly common phenomenon in the age of digital, interactive, and mobile technologies. The French sociologist Gabriel de Tarde, forefather of Diffusion of Innovations Theory, described this phenomenon using the metaphor of waves changing their shape while spreading from their origin.

Nevertheless, this phenomenon has largely been neglected in Diffusion of Innovations Theory. From the seminal study undertaken by Ryan and Gross (1943) regarding the diffusion of hybrid seed corn, through the subsequent editions of *Diffusion of Innovations* (Rogers, 1962, 1983, 1995, 2003; Rogers & Shoemaker, 1972), the qualitative evolution of innovations has rarely been the focus of interest.

Only during the 1970s did the concept of Re-Invention (e.g., Charters & Pellegrin, 1973; Glick & Hays, 1991; Goodman & Steckler, 1989) enter the field of Diffusion of Innovations Theory, analyzing the circumstances leading to a change in the innovation during the diffusion process. However, until today, these ideas have not led to a change in the theoretical basis of diffusion theory, which remained attached to a linear process, based on the binary logic of adoption versus rejection. As we argued above, this lack of theoretical progress is due in part to the transmission model of communication, which remains fundamental to diffusion theory today. Consequently, the way of overcoming this limitation lies in following theoretical approaches from outside Diffusion of Innovations Theory, which are based on a less linear understanding of communication. In outlining the central ideas of Cultural Studies (Hall, 1980; Silverstone & Haddon, 1996), Social Learning Theory (Bandura, 1977), and the Mobile Phone Appropriation Model (Wirth, von Pape & Karnowski, 2008), we elaborated upon two ideas that are crucial for a broader concept of adoption.

First, the creativeness of adopters during the diffusion process must be taken into account. This point is illustrated by the idea of negotiating meaning during this process, taken from Cultural Studies. In addition, Social Learning Theory conceptualizes the imitation of modeled behavior as a creative act, tailoring the imitation of the modeled behavior to the specific circumstances. The MPA model models this idea and differentiates the original binary decision (adoption versus rejection) to a variety of possible forms of actually using an innovation, thereby addressing Rogers' (2003, p. 181) suggestion to investigate "adoption at the implementation stage, as change that has really happened."

Second, this user creativity in the diffusion process is powered by meta-communication. Here, Cultural Studies emphasizes the importance of communication between either different potential adopters or adopters and inventors: the negotiation of meaning. Social Learning Theory focuses mainly on the effects of observing and imitating during the diffusion of a certain form of behavior, although also taking communication into account, e.g., in the form of reinforcement. The MPA model molds meta-communication (either interpersonal or mass-mediated communication and observation) to power the whole process. These ideas of meta-communication go along perfectly with Rogers' (2003, p. 5) definition of diffusion being a communication process: "Diffusion is the process in which an innovation is *communicated* [emphasis added] through certain channels over time among the members of a social system." Thus, meta-communication has always been at the heart of Diffusion of Innovations Theory, although few studies have incorporated the aspect of communication during the diffusion process, confirming the influence of mass and interpersonal communication on the diffusion process. Weber and Evans (2002) found a positive relationship between media coverage of a media innovation and its diffusion. The influence of interpersonal communication on the diffusion of media innovations (e.g., the Internet, cell phones, Wi-Fi services) has also been confirmed in previous studies (see Lagos, 2008; Rhee & Kim, 2004; Wei, 2001, 2006; Zhu & He, 2002).

Media innovations nowadays are highly prone to Re-Invention, according to the factors brought up by Re-Invention research in recent years. Consequently, we argue that Diffusion of Innovations Theory will gain momentum in analyzing the diffusion of recent media innovations, as well as other innovations that are highly susceptible to Re-Invention, when seeking to overcome the linear structure of the diffusion process that is inherent in the binary logic of adoption.

References

Ajzen, I. (1985). From intentions to actions: A theory of planned behavior. In J. Kuhl & J. Beckmann (Eds.), *Springer series in social psychology: Action control. From cognition to behavior* (pp. 11–39). Berlin: Springer.

Ajzen, I. (2005). *Attitudes, personality and behavior* (2. ed.). Maidenhead, UK: Open University Press.

Bandura, A. (1977). *Social learning theory*. Englewood Cliffs, NJ: Prentice Hall.

Bandura, A. (1986). *Social foundations of thought and action: A social cognitive theory*. Englewood Cliffs, NJ: Prentice Hall.

Bandura, A. (2001). Social cognitive theory of mass communication. *Media Psychology*, *3*(3), 265–299.

Bandura, A. (2006). On integrating social cognitive and social diffusion theories. In A. Singhal & J. W. Dearing (Eds.), *Communication of innovations: A journey with Ev Rogers* (pp. 111–135). New Delhi, Thousand Oaks, London: Sage.

Brown, M. Ellen (1994). *Soap opera and women's talk: The pleasure of resistance*. Thousand Oaks, CA: Sage.

Certeau, M. de (2002). *Arts de faire* (Nouvelle éd., Vol. 146). Paris: Gallimard.

Charters, W. W., Jr., & Pellegrin, R. J. (1973). Barriers to the innovation process: Four case studies of differentiated staffing. *Educational Administration Quarterly*, 9(1), 3–14.

Coleman, J., Katz, E., & Menzel, H. (1957). The diffusion of an innovation among physicians. *Sociometry*, *20*(4), 253–270.

Davis, F. D. (1986). *A technology acceptance model for empirically testing new end-user information systems: Theory and results*. Boston: MIT Press.

Dearing, J. W., & Singhal, A. (2006). Communication of Innovation: A Journey with Ev Rogers. In A. Singhal & J. W. Dearing (Eds.), *Communication of innovations: A journey with Ev Rogers* (pp. 15–28). New Delhi, Thousand Oaks, London: Sage.

Dearing, J. W., & Meyer, G. (2006). Revisiting diffusion theory. In A. Singhal & J. W. Dearing (Eds.), *Communication of innovations: A journey with Ev Rogers* (pp. 29–60). New Delhi, Thousand Oaks, London: Sage.

Donner, J. (2007). The rules of beeping: exchanging messages via intentional "missed calls" on mobile phones. *Journal of Computer-Mediated Communication, 13*(1).

Downs, G. W. (1976). *Bureaucracy, innovation, and public policy*. Lexington, MA: Lexington.

Fischer, C. S. (1992). *America calling: A social history of the telephone to 1940*. Berkeley, CA: University of California Press.

Fishbein, M., & Ajzen, I. (1975). *Believe, attitude, intention and behavior: An introduction to theory and research*. Reading, MA: Addison-Wesley.

Frissen, V. (2000). ICTs in the rush of life. *The Information Society*, 16(1), 65–75.

Glick, H. R., & Hays, S. P. (1991). Innovation and re-invention in state policymaking: Theory and the evolution of living will laws. *Journal of Politics*, *53*(3), 835–850.

Goffman, E. (1974). *Frame analysis: An essay on the organization of experience*. New York: Harper & Row.

Goodman, R. M., & Steckler, A. (1989). A model for the institutionalization of health promotion programs. *Family & Community Health, 11*(4), 63–78.

Hall, S. (1980). Encoding/Decoding. In S. Hall; D. Hobson, & A. Lowe (Eds.), *Culture, media, language: Working papers in cultural studies; 1972–79* (pp. 128–138). London: Routledge.

Hall, S. (2003). Encoding/Decoding. In M. G. Durham & D. M. Kellner (Eds.), *Media and cultural studies: Key works* (pp. 166–176). Malden, MA: Blackwell.

Havelock, R. G. (1974). Locals say innovation is local: A national survey of school superintendents. In S. Temkin & M. V. Brown (Eds.), *What Do Research Findings Say About Getting Innovations Into Schools: A Symposium* (pp. 73–95). Philadelphia.

Hays, S. P. (1996a). Influences on re-invention during the diffusion of innovations. *Political Research Quarterly*, 49(3), 631–650.

Hays, S. P. (1996b). Patterns of re-invention: The nature of evolution during policy diffusion. *Policy Studies Journal*, *24*(4), 551–566.

Höflich, J. R., & Rössler, P. (2001). Mobile schriftliche Kommunikation oder: E-Mail für das Handy. *Medien & Kommunikationswissenschaft*, 49(4), 437–461.

Johnson, B., & Rice, R. E. (1984). Re-invention in the innovation process: The case of word processing. In R. E. Rice and associates (Eds.), *The new media: Communication, research and technology* (pp. 157–184). Beverly Hills, CA: Sage.

Jonas, K., & Doll, J. (1996). Eine kritische Bewertung der Theorie überlegten Handelns und der

Theorie geplanten Verhaltens. *Zeitschrift für Sozialpsychologie*, *27*(1), 18–31.

Karnowski, Veronika & von Pape, Thilo (2009). MobileTV im Alltag der Nutzer. Ergebnisse einer dreimonatigen Panelstudie. In J. Krone (Ed.), *Fernsehen im Wandel. Mobile Tv & IPTV in Deutschland und Österreich* (pp. 241–255). Baden-Baden, Germany: Nomos.

Katz, E. (1999). Theorizing diffusion: Tarde and Sorokin revisited. *The ANNALS of the American Academy of Political and Social Science*, 566(1), 144–155.

Katz, E., Blumler, J. G., & Gurevitch, M. (1974). Utilization of mass communication by the individual. In J. G. Blumler & E. Katz (Eds.), *Sage annual reviews of communication research: Vol. 3. The uses of mass communications. Current perspectives on gratifications research*. Beverly Hills, CA: Sage.

Kendzierski, D. (1990). Decision making versus decision implementation: An action control approach to exercise adoption and adherence. *Journal of Applied Social Psychology*, *20*(1), 27–45.

Lagos, T. G. (2008). Mediating commons: Rural Greece. *New Media Society*, *10*(4), 565–583.

Lasswell, H. (1948). The structure and function of communication in society. In L. Bryson (Ed.), *The communication of ideas* (pp. 37–51). New York: Cooper Square.

Lehtonen, T. (2003). The domestication of new technologies as a set of trials. *Journal of Consumer Culture*, 3(3), 363–385.

Leung, L., & Wei, R. (2000). More than just talk on the move: Uses and gratifications of the cellular phone. *Journalism & Mass Communication Quarterly*, *77*(2), 308–320.

Lewis, L. K., & Seibold, D. R. (1996). Communication during intraorganizational innovation adoption: Predicting users' behavioral coping responses to innovations in organizations. *Communication Monographs*, 63(2), 131–157.

Ling, R. (2004). *The mobile connection: The cell phone's impact on society*. San Francisco: Morgan Kaufmann.

Ling, R., Nilsen, S., & Granhaug, S. (1999). The domestication of video-on-demand: Folk understanding of a new technology. *New Media Society*, *1*(1), 83–100.

Lowery, S. A., & DeFleur, M. L. (Eds.). (1995). *Milestones in mass communication research: Media effects* (3. ed.). White Plains, NY: Longman.

Majchrzak, A., Rice, R. E., Malhotra, A., King, N. & Ba, S. (2000). Technology adaptation: The case of a computer-supported inter-organizational virtual team. *MIS Quarterly*, *24*(4), 569–600.

Menzel, H., & Katz, E. (1955). Social relations and innovation in the medical profession: The epidemiology of a new drug. *Public Opinion Quarterly*, *19*(4), 337–352.

Meyer, G. (2004). Diffusion methodology: Time to innovate? *Journal of Health Communication*, 9, 59–69.

Oksman, V., & Turtiainen, J. (2004). Mobile communication as a social stage: Meanings of mobile communication in everyday life among teenagers in Finland. *New Media Society*, 6(3), 319–339.

Orlikowski, B. (1993). CASE tools as organizational change: Investigating incremental and radical changes in systems development. *MIS Quarterly*, *17*(3), 309–340.

Ouellette, J. A., & Wood, W. (1998). Habit and intention in everyday life: The multiple processes by which past behavior predicts future behavior. *Psychological Bulletin*, *124*(1), 54–74.

Quandt, T. & von Pape, T. (2010), Living in the mediatope: a multi-method study on the evolution of media technologies in the domestic environment. *The Information Society*, *26*(5), 330–345.

Rhee, K. Yong, & Kim, W. B. (2004). The adoption and use of the Internet in South Korea. *Journal of Computer-Mediated Communication*, 9(4).

Rice, R. E., & Rogers, E. M. (1980). Re-Invention in the innovation process. *Knowledge: Creation, Diffusion, Utilization*, *1*(4), 499–514.

Rogers, E. M. (1962). *Diffusion of innovations*. New York: Free Press.

Rogers, E. M. (1983). *Diffusion of innovations* (3. ed.). New York: Free Press.

Rogers, E. M. (1995). *Diffusion of innovations* (4. ed). New York: Free Press.

Rogers, E. M. (2003). *Diffusion of innovations* (5. ed). New York: Free Press.

Rogers, E. M., & Shoemaker, F. (1972). *Communication of innovations* (2. ed). New York: Free Press.

Rogers, E. M, Vaughan, P. W., Swalehe, R. M. A., Rao, N., Svenkerud, P., & Sood, S. (1999). Effects of an entertainment-education radio soap opera on family planning behavior in Tanzania. *Studies in Family Planning*, *30*(3), 193–211.

Ryan, B., & Gross, N. C. (1943). The diffusion of hybrid seed corn in two Iowa communities. *Rural Sociology*, 8(1), 15–24.

Schenk, M., Dahm, H., & Sonje, D. (1996). *Innovationen im Kommunikationssystem: Eine empirische Studie zur Diffusion von Datenfernübertragung und Mobilfunk*. Münster: LIT.

Silverstone, R., & Haddon, L. (1996). Design and the domestication of information and communication technologies: Technical change and everyday life. In R. Mansell & R. Silverstone (Eds.), *Communication by design: The politics of information and communication technologies* (pp. 44–74). Oxford: Oxford University Press.

Singhal, A., Cody, M. J., Rogers, E. M., & Sabido, M. (2004). *Entertainment-education and social change: History, research, and practice*. Mahwah, NJ: Lawrence Erlbaum.

Tarde, G. (1890). *Les lois de l'imitation: étude sociologique*. Paris: Alcan.

Tarde, G. (1902). L'invention considérée comme moteur de l'évolution sociale. *Revue Internationale de Sociologie*, *10*(7), 561–574.

Valente, T. W. (2006). Communication network analysis and the diffusion of innovations. In A. Singhal & J. W. Dearing (Eds.), *Communication of innovations: A journey with Ev Rogers* (pp. 61–82). New Delhi, Thousand Oaks, London: Sage.

Vaughan, P. W., & Rogers, E. M. (2000). A staged model of communication effects: Evidence from an entertainment-education radio soap opera in Tanzania. *Journal of Health Communication*, 5(3), 203.

von Hippel, E. (1986). Lead User: A source of Novel Product Conepts. *Management Science*, *32*(7), 791–805.

von Pape, T., Karnowski, V. & Wirth, W. (2008). Die Mobile Phone Appropriation-Skala (MPA-Skala): Konstruktion und Evaluation. In Jörg Matthes, Werner Wirth, Andreas Fahr & Gregor Daschmann (Eds.), *Die Brücke zwischen Theorie und Empirie: Operationalisierung, Messung und Validierung in der Kommunikationswissenschaft* (pp. 96–127). Köln, Germany: Halem Verlag.

von Pape, T., Karnowski, V., Wirth, W., Klimmt, C. & Hartmann, T. (2007, May). *Living in*

Oblivion—a diary-study on the appropriation of a role-playing game. Paper presented at the 57th Annual Conference of the International Communication Association (ICA), San Francisco, May 24–28, 2007.

Weber, I. A. N., & Evans, V. (2002). Constructing the meaning of digital television in Britain, the United States and Australia. *New Media Society*, *4*(4), 435–456.

Wei, R. (2001). From luxury to utility: A longitudinal analysis of cell phone lagards. *Journalism & Mass Communication Quarterly*, *78*(4), 702–719.

Wei, R. (2006). Wi-Fi powered WLAN: When built, who will use it? Exploring predictors of wireless Internet adoption in the workplace. *Journal of Computer-Mediated Communication*, *12*(1), 155–175.

Wirth, W., von Pape, T., & Karnowski, V. (2008). An integrative model of mobile phone appropriation. *Journal of Computer-Mediated Communication*, *13*(3), 593–617.

Wissler, C. (1914). The influence of the horse in the development of plains culture. *American Anthropologist*, *16*(1), 1–25.

Zhu, J. J. H., & He, Z. (2002). Diffusion, use and impact of the Internet in Hong Kong: A chain process model. *Journal of Computer-Mediated Communication*, *7*(2).

· 4 ·

Theoretical and Methodological Starting Points for a Situational Approach towards the Understanding of the Adoption and Use of New Technologies and Services

Lidwien van de Wijngaert & Harry Bouwman

Introduction

This chapter discusses theories that help to understand the adoption and use of new technologies and services. We will also discuss the shortcomings of these theories as well as an incitement of an alternative approach. We conclude that, in order to understand the adoption and use of new technologies, we need to study the interaction between the emergence of needs and the use of services and how time influences this process. We argue that individual, situational, technology, and contextual factors influence the process. In the second half of the chapter we propose a research method that can help to empirically test this alternative approach. Using the basics from factorial survey, vignette studies, and conjoint measurement, we will discuss a research method that closely connects to the presented theoretical framework. The basic idea is to present people with contrived hypothetical situations. These situations, scenarios, vignettes, or cases are developed by combining characteristics of possible situations. Together with individual and contextual characteristics the situational characteristics can be used to explain possible future use of new technologies. We will discuss the design issues of such a study as well as the data-analysis issues that have to be dealt with.

Taking on the User Perspective

Most research projects that are related to understanding the adoption and use of technologies and services focus on technical aspects such as safe transactions, embedded systems, and service-oriented architectures. The perspective of the technology and/or supplier is central to these projects. Projects in which social scientists participate often focus on issues such as Human Computer Interaction and Usability. Although the user plays a role in this latter type of research, the perspective still is that of the supplier of technology and organizations that need to know how a service can best be offered. A much more limited amount of research is aimed at understanding the actual adoption and use of technologies and services from a user perspective. Effects are hardly studied at all. Although many technologies and services are developed, the actual use of these services is often lagging behind. More and more, it is becoming clear that the technical and organizational issues are not the bottleneck for the adoption and use of services. In order to understand why people are using services (or maybe even why they are not using them), we need to take a user perspective. However, when push comes to shove, research is still strongly aimed at the technological and organizational issues. Even in research in which the goal is to measure the adoption and use of, e.g., services, the offline alternatives are often not taken into account. Respondents are asked to describe their online behaviors but not the context in which they usually choose between an on- or offline alternative.

Current Theoretical Models

In this chapter we shift the focus to the user perspective. As a starting point for this contextual approach, we make use of Rogers' Diffusion of Innovations and the Technology Acceptance Model (TAM). Rogers (2003) has provided a broad set of insights into the diffusion of innovations. His research covers the macro as well as the micro level and the demand side as well as the supply side. With Rogers we share the opinion that it is important to look at all aspects of an innovation in order to understand how a new technology will develop in society. In the end, Rogers has provided us with a broad terminology to describe the diffusion on innovations. Terms such as Innovators and Early adopters alongside terms such as S-curve have become common not only in diffusion research but even in day-to-day life. The five core concepts used in Diffusion of Innovations literature are relative advantage, complexity, compatibility, triability, and observability (Rogers, 1995). Similar concepts play a role in the widely used Technology Acceptance Model (TAM) (Davis, 1989). This model intends to explain behavioral intention to use technologies based on the Theory of Reasoned Action (Fishbein & Ajzen, 1975) and Theory of Planned Behavior (Ajzen, 1991). Central concepts are perceived ease of use and perceived usefulness (Davis, 1989).

Despite the many insights that diffusion research has provided, criticism is also possible. This criticism is described very well by a special issue on the Technology Acceptance Model (TAM) and the Unified Model of Technology Adoption and Use (UTAUT) (Venkatesh, Morris, Davis & Davis, 2003) that was published in the *Journal of the Association of Information Systems* in April 2007. The special issue sketches the strong and weak points of this theoretical model. Benbasat and Barki argue that although TAM is one of the most influential theories in Information Systems, the theory is in a state of theoretical chaos and confusion (Benbasat & Barki, 2007; Straub & Burton-Jones, 2007). With this proposal, we intend to join the criticism with regard to current theoretical models. At the same time, we offer a framework that will help to better understand the adoption, use, and effects of services.

Shortcomings of Current Models

To begin with, in most studies, actual use is not measured. Most studies focus on behavioral intention. Moreover, in cases where actual use is measured, it is not measured on the basis of log-data, but on the basis of self-reports, which means that what is measured is not the actual use, but rather self-reported use. Second, in many cases the research population consists of students. Student-based studies are found to offer stronger support for a TAM-based approach than studies involving other kinds of populations (Schepers & Wetzels, 2007). A third criticism is that it is unclear to what extent TAM actually provides insight into (intended) user behavior. In recent TAM studies, the focus is on concepts such as self-efficacy (Mao, Srite, Thatcher & Yaprak, 2005), computer experience (Kleijnen, De Ruyter & Wetzels, 2004), subjective norms (Legris, Ingham & Collerette, 2003; Venkatesh & Davis, 2000), hedonistic values (Van der Heijden, 2003), perceived enjoyment (Cyr, Head & Ivanov, 2006), and personal innovativeness (Lu, Yu, Liu & Yao, 2003). However, the fundamental problem with TAM is that, arguably, the intention to use information technology is (too) closely related to its perceived usefulness. In the original TRA approach (Fishbein & Ajzen, 1975), beliefs and behavioral intention are interrelated. In TAM research, the focus is on one of these beliefs, e.g., perceived usefulness, which is related to intention to use. This may suggest that these are two concurrent measurements of the same construct. To put it more formally, although in statistical terms there is discriminant validity, we observe a strong convergence between the two constructs. The comparable formulations of the items that belong to the different scales invoke high correlations between the scales. The high levels of explained variance in many TAM papers can be attributed to the strong relationship between the two concepts.

In sum, TAM treats technology as a black box, rather than taking specific characteristics of technology that may influence people's (intention to) use (of) a technology into account (Amberg, Hirschmeier & Wehrmann, 2004; Legris et al., 2003; Lu et al., 2003). The question that needs to be answered is: which kinds of user needs, preferences, behaviors, and contexts are related to the use of which types of on- and offline services.

Alternative Approaches

Rather than picking a single theoretical framework to look at the adoption and use of a new technology, Monge and Contractor (2003) propose that a wide variety of theories and levels are needed to understand the complexity at hand.

An alternative approach to understanding the adoption and use of online services is based on the Adaptive Structuration Theory (AST) (DeSanctis & Poole, 1994; Poole & DeSanctis, 1990). AST is based on the notion that task-related characteristics need to match technological characteristics (and vice versa) if effective communication is to take place. The concept of fit was introduced by Daft and Lengel (1984, 1986) and later extended by Goodhue and Thompson (1995) and Dishawa and Strong (1999). Their approaches are known as the media choice, media richness theory, and the task-technology-fit approach. The media choice, media richness theory suggests that not all information and communication technologies are equally suited to meet the information requirements associated with specific tasks. The media choice, media richness theory also has been the subject of considerable criticism, especially with regard to some of the underlying assumptions (Fulk, Schmitz & Steinfield, 1990): communication technologies and tasks are assumed to have fixed characteristics; all users are supposed to be aware of the inherent differences between the various available technologies and tasks; the social environment is assumed to have no influence on the choice process; and media choice is viewed as a rational and cognitive process motivated solely by efficiency.

More subjective approaches, such as the Social Influence Model (Schmitz & Fulk, 1991), the Channel Expansion Theory (Carlson & Zmud, 1997) and Social Information Processing theory (Walther, 1992), are based on the assumption that the characteristics of task and medium cannot be determined objectively. Additionally, Webster and Trevino (1995) propose a combined method. The same applies to matching information and communication technology and task: different users will make different choices based on their individual characteristics. Other, more qualitative approaches that can be of help are social shaping and social construction of technology (Bijker, 1997; Lievrouw & Livingstone, 2002; Williams & Edge, 1996), domestication (Frissen, 2000; Mante-Meijer & Haddon, 2001; McCreadie & Rice, 1999a, 1999b; Silverstone & Haddon, 1996), and critical mass theory (Markus, 1990; Olson, 1965).

To summarize the above, we distinguish a number of central themes that we need to understand in order to describe the adoption use and effect of public electronic services:

- How the process of adoption and use of services takes place
- How the interaction between the emergence of needs and the use of services works
- How time influences these processes
- How individual, situational and technology factors influence the adoption and use of services
- How contextual factors on the meso and macro level of society influence the process
- What the effects of using services are on the micro, meso and macro levels

Towards a Contextual and Process Approach

We propose to integrate the themes that were derived from the theoretical notions listed above by using a process model as presented in Figure 1. In the center of the model we see how needs and the use of services alternate. The actual use of a service is invoked by a need, which may be the consequence of a specific individual in a specific situation. A need may also be the result of the use of a service. Beside this day-to-day circle of needs and use, there is a more long-term circle of adoption and rejection of services.

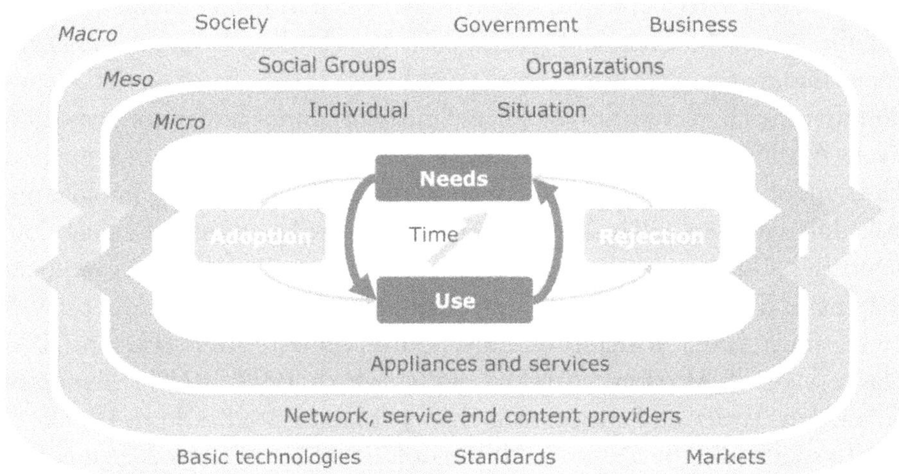

Figure 1. Point of Departure: A Process Model of the Use of E-services.

By using a process model as a starting point, we intend to stress that change over time is inherent to the adoption, use, and effects of services. If we look at the layers around the process, we see that many factors may influence the process. On the micro level, closest to the user, we see that individual, situational, and technology characteristics influence the process. These factors are assumed to have the most strong and direct influence on user behavior. On the meso and macro level we see organizations, (technological) trends, social and cultural context, and society influence the process. The interlacing of the layers suggests that the different layers can have an effect not only on the process, but also on each other. Similar attempts to develop a general framework have been developed by Heres, Mante-Meijer, Turk, and Pierson (2005), Blechar, Knutsen, and Damsgaard (2005), and Ciaborra and Willocks (2006). These frameworks also attempt to integrate the different concepts into a single framework. However, none of these frameworks have been tested empirically. In the rest of this chapter we would like to provide some suggestions on how this empirical test can be developed. As became clear in the theoretical framework, there are two quite separate schools in diffusion research: a more quantitative approach and a more qualitative approach. Also, we advocate a contextual approach in which several levels can be distinguished. Consequently, like Monge et al.

(2003), we intend to integrate qualitative and quantitative approaches in a multi-theory, multi-level approach.

Empirically Testing the Framework

We suggest bringing the realms of qualitative and quantitative research together by researching the subject of technology diffusion and use side by side. In the qualitative approach the focus of the research is on understanding Figure 1 by closely examining the user in his or her context. By taking up a holistic approach, the research will obtain insight into the ins and outs of the adoption, use and effects of public services. The research methods that can be used are, e.g., interviews, diaries, and focus groups. Diary research has been used in several research areas, such as medical and mobility studies. The use of diaries for (information) technology is relatively new. In a diary research project respondents essentially keep track of their daily activities. The advantage of the diary method is that people do not have to rely on their (selective) memory. Consequently, a more complete and correct view on daily usage can be obtained through diary research. Essentially, a combination of web statistics or log files and diaries can provide in-depth insight into how people use technologies. When performing a diary study many choices need to be made. For example, choices regarding the use of solicited versus unsolicited and a choice for a full activity diary versus a total time diary. A full description of all choices that need to be made and possible results can be found in Vermaas and Van de Wijngaert (2005). The results of this type of research have value in itself, but can also directly feed into the quantitative approach.

A second, more quantitative towards the understanding of the adoption and use of new technologies is the Social Network approach. A Social Network study can help to assess the (complex) networked structure of the decision to adopt technologies and services. A network study can provide insight into the information flows between service providers and users and among users themselves and the channels they use. The analysis of these data can provide insight into the social structures in which communication takes place as well as in the contextual determinants that trigger and shape the information flows. By using the Social Network approach, in-depth insight into, for example, social influence can be studied in detail.

The third research method, noted earlier, has a stronger quantitative nature and is referred to as the vignette approach towards understanding the adoption and use of new technologies and services. In this approach, a model with a limited number of factors that can explain the adoption and use of technology and services is developed and tested empirically. An important element of this approach is the inclusion of contextual factors. In the following sections, we will describe three research approaches.

Although many examples of research that are related to the framework could be presented here, we limit ourselves to the vignette approach. We will provide a short description of the method and an example of research in the area of e-government.

Vignette Studies in Practice

The fundamental problem of understanding the adoption and use of new information and communication technologies is that potential users are not familiar with them. Consequently, it is difficult to ask people if and how they would use these services and applications.

This section discusses a methodology that makes it possible to judge the suitability of the use of new information and communication services while taking into account characteristics of the technology, characteristics of the context of the user, and user characteristics. Using the basics from factorial survey, vignette studies, and conjoint measurement, we will discuss the design issues of such a study as well as the data-analysis issues that have to be dealt with.

The possible success of a new technology or service largely depends on the existence of a (latent) need for that technology or service. However, research in customer needs for new services that are not readily available is hindered by the fact that respondents do not have any (or only a vague) idea about the specific characteristics of the technology or service. It becomes even more difficult to predict the possible success when new services are not only improvements of existing ones, but offer completely new and non-existing concepts of information, communication, entertainment, or transaction. Moreover, in many cases adoption does not only entail the adoption of a single technology but adoption is concerned with a bundle of technologies.

In this section we propose the use of a research method that is able to deal with this complexity by combining the strengths of survey and experimental research. As a basis, we will first describe the similarities and differences of the roots of this method: policy capturing, vignette studies, and conjoint measurement. Using our experience from a variety of studies, we will then describe the central steps, issues, and design choices that need to be taken when conducting such a study. Next, we will discuss the design and results of a study that we conducted earlier. Although this paper has a strong focus on research design, we will draw a conclusion with regard to the development of a conceptual framework for understanding the success of technologies or services.

Research Methods Considered

The research method that is central in this paper is also referred to as factorial surveys, vignette studies, policy capturing, and conjoint measurement. These methods offer similar yet more or less unrelated approaches in marketing and product development research and have been around for some time. Conjoint measurement (Molin, 1999) is especially popular in marketing and product design. Policy capturing is popular in organizational studies, whereas vignette studies can mostly be found in the domain of, e.g., health care. Although strongly related it appears that the methods are developed independently from each other. All three methods borrow and adapt the concept of manipulation from the experimental tradition. From the survey tradition they borrow the greater richness of detail and complexity that characterizes real-life circumstances (Rossi & Nock, 1982). The basic idea is to present people with contrived hypothetical situations. These situations, scenarios, vignettes, or cases are developed by combining characteristics of possible situations.

In order to introduce the method as well as show its broad applicability, we will first describe the roots on which it is built. Factorial survey (Rossi et al., 1982), also known as vignette study, is probably the oldest method in this range. The first trials date from 1951. Factorial survey is a method "for studying human choice when the alternatives are complex bundles that vary in many ways" (Rossi et al., 1982, p. 10). The factorial design approach is used in studies in a variety of topics such as social status, family prestige, sexual harassment, child abuse, political norms, and justice judgments, but factorial survey is also used in studies to assess the willingness to pay for specific services. Policy capturing is a method employed by researchers to assess how decision makers use available information when making evaluative judgments (Karren & Woodard Barringer, 2002). Basically the research is directed to the analysis of how individual decision makers weigh, combine, or integrate information (Zedeck, 1977, p. 51). Policy capturing is mainly used in organizational studies, for instance, discussing media choice of managers, applicant screening, organizational downsizing, and performance appraisal. Conjoint measurement (Vriens,

1995; Molin, 1999) is a classical technique to identify consumer preferences in a multi-attribute decision-making. The origin of conjoint measurement is assigned to the psychologist Luce and the statistician Tukey, according to Gustafsson et al. (2003).

Since the introduction of marketing research in the 1970s, such research has maintained attention from theoreticians and practitioners. The number of studies exceeds an estimated 4000. Conjoint measurement has been used for new product planning, improvement of existing achievements, pricing policies, advertising, distribution, market segmentation, and simulation of purchasing decisions.

Designing a Study

When conducting a study using this method, researchers have to make a number of choices. Based on the review of research by Karren et al. (2002) and Gustafsson et al. (2003), we discuss best practices and issues as realism of the scenarios, the limits of a full factorial design, need for orthogonal cues, sample size and statistical power, and the assessment of reliability. This section systematically describes the choices, alternatives, and the pros and cons of different solutions.

Creation of the Cases

The starting point of any study that uses this method is the dependent variable: adoption and use of technologies and services. As outlined in the previous section this variable is explained by a large number of contextual (i.e., "traditional" variables such as organizational and individual characteristics. In addition, we propose to add a number of situational characteristics). On a macro level these can be policy measures (such as subsidies or laws); on a meso level these can be social influence or service characteristics such as price and user friendliness. On a micro level we distinguish situational characteristics such as urgency and mobility. Figure 2 provides some examples of such variables.

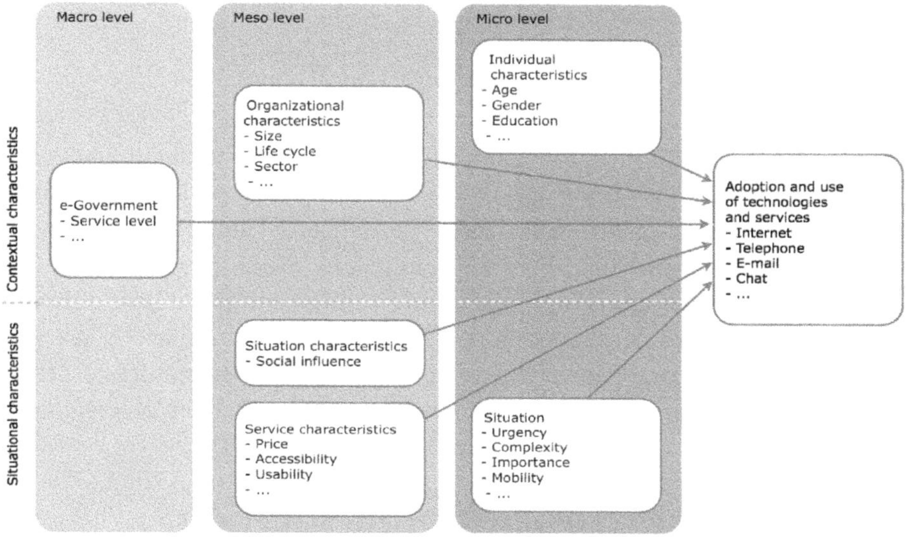

Figure 2. A General Conceptual Model of Technology Adoption and Use.

A case is a description of a situation characterized by a unique combination of values on two or more variables. In factorial surveys the set of all possible combinations of all values and variables is called the factorial object universe.

A simple calculation demonstrates that only a few variables (or dimensions), with a few values (or levels), can already create a large number of objects. To use an example of an object—for instance, a mobile device—the relevant attributes might be display size, price, technical capabilities, et cetera. The levels of these attributes might be, for instance, price, price-classes between €100 and €200, €201 and €300, and more than €300. The universe of all factorial objects is based on all the relevant attributes and their levels. So if there are three relevant attribute sizes (traditional GSM-display, small display, or PDA-format display), price (100 and 200 €, 201 and 300 € and more than 300 €), and technical capabilities (9.6 kbps, 144 kbps, 2 Mb), with each three value-levels, the universe will consist of 27 unique objects. We may assume, however, that the number of relevant attributes is normally much higher. The size of factorial objects can significantly increase by adding an attribute with multiple levels, for instance, ten. More formally expressed, a few dimensions (k), with a few levels (q), can already create a large number of objects. In the above example, we see that for $k1$ (q=3), $k2$ (q=3), and for $k3$ (q=3). So according to [1], the number of objects in the factorial object universe is $(3) \cdot (3) \cdot (3) = 27$.

$$\prod_{i=1}^{k} q_i \qquad \text{[Equation 1]}$$

It is clear that with increasing numbers of attributes and an increasing number of levels for each attribute, the number of alternatives will increase rapidly. A respondent cannot deal with these large numbers of alternatives (stimuli). All kind of effects—tiredness, boredom, et cetera—will influence the validity of the research. Factorial design and conjoint measurement follow different strategies in dealing with the large number of potential alternatives. We will first discuss the approach taken in factorial design and policy capturing and then discuss the way conjoint measurement tries to solve the problem of large numbers.

The Factorial Design Solution: A Sample of Cases

In factorial design studies many dimensions and/or levels resulting in a large number of cases in the factorial object universe are object of study. By getting away from the assumption that in a full factorial design all vignettes have to be rated by every individual respondent, towards an approach in which multiple respondents can rate a limited number of vignettes, the applicability of vignette studies was increased. Therefore, the factorial object sample was introduced: a random sample of all possible scenarios. The factorial object sample is an unbiased sample of the objects in a factorial object universe. This sample consists of all cases that are included in the actual research and assumed to be representative for all cases. Most of the time the number of scenarios is still quite numerous and cannot be evaluated by every single respondent. Respondents are presented a sub-sample from the factorial object sample. The respondent sub-sample is an unbiased sample of a factorial object sample that is given to a single respondent for judgment. The relation between the factorial object universe, the factorial object sample, and the respondent sub-sample is depicted in Figure 3.

An unbiased sample of potential factorial objects is used. In the situation in which data is collected, not all factorial objects will be presented to the respondent. The respondents sub-sample is an unbiased sample of the universe of all factorial objects and is randomly presented to the respondent. More than one factorial object is judged by a respondent, multilevel problems (intra-respondent correlation, Rossi & Anderson, 1982, p. 33) become an issue and have to be dealt with.

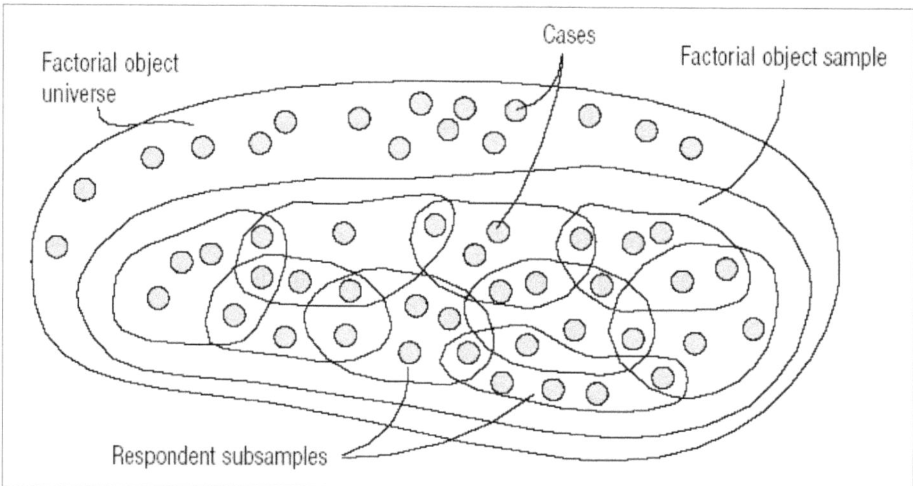

Figure 3. Factorial Object Universe, Object Sample, and Respondent Sub-Sample.

The Policy Capturing Solution: Confounded Factorial Design

Policy capturing takes another approach to deal with the potential large number of scenarios that has to be dealt with. Respondents only read a subset of the full factorial set. Basically the researchers follow approaches traditionally used in experimental research design. These approaches are known as confounded factorial design, i.e., incomplete block design and fractional factorial design (Karren et al., 2002, pp. 348–349). Cue sets are presented in such an orthogonal way that the assessment of each variable's independent contribution can be assessed but not higher order interaction effects. Cue sets for the incomplete block design and the factorial design are created in a similar way. In both, the full factorial design set is divided into blocks, and each respondent is presented one of the blocks. Each block is composed of a unique set of scenarios and is created by dividing the full set into halves, quarters, eights, et cetera. The difference between the incomplete block design and the fractional factorial design is that in the incomplete block design all of the subsets are used, while in the factorial design only one subset is used. In practice, this means that in an incomplete block design all subgroups of participants receive a different subset, while in a fractional design all participants receive the same subset of scenarios. In an incomplete block design the number of respondents is dependent on the number of blocks used, for example, in a full design one respondent judges all the scenarios, in a two-block design half of the scenarios are judged by two respondents, in a four-block design a quarter of the full design are judged by four respondents, et cetera.

Research in the validity of confounded factorial design compared with a full factorial design is scarce. The only study (Graham & Cable, 2001) indicates that effects of independent variables were comparable when comparing a full factorial design and a confounded design. Nevertheless, Karren and Woodard Barringer (2002) suggest that the use of full factorial design is more prudent when possible.

The Conjoint Measurement Solution: Fewer Variables and Levels

Although conjoint experts prefer the profile method or multiple factor evaluation as being most realistic, conjoint measurement is limited to approximately eight attributes, according to Gustafsson et al. (2003). The full factor approach is comparable with the approach taken in factorial survey and policy capturing. Conjoint measurement differs here from the other two methods mainly, based on selection steps, because only a limited number of alternatives is being analyzed. This basically results into looking at the effect of single factors individually or two-factors. The single factor analysis, which conflicts with the general notion of conjoint measurement, i.e., looking into the decomposition of a decision in a number of factors, is basically a result of the opportunities offered by the hybrid or adaptive conjoint measurement approach.

In a two-factor approach, matrices are used in which two factors, the basis for an evaluation of the possible combinations between the different levels of the attributes, are used. The results are a ranking of alternatives and are therefore limited as far as data-analysis is concerned. This is an example of the reduced design: two attributes are presented at a time and respondents are asked to rank-order the full set of combinations for each pair. It is less complex than a full-profile approach and less stressful; moreover, the information overload for respondents is less. The downside of this approach is that a larger number of respondents is required.

The hybrid or adaptive conjoint measurement approach (ACA) involves a two-stage procedure in which respondents are first asked to rate the desirability of the full set of factors. The factors vary according to individual respondents' desirability ratings. This approach is supported by software that allows researchers to generate individualized scenarios. In principle, ACA is a more modern version of the two-factor approach.

Summarizing, there are two approaches to limiting the number of scenarios that are presented to respondents: either by limiting the number of possible scenarios or by drawing a sample from a large number of possible cases. Our preference is to limit the number of cases. The first reason for this is that we prefer a theoretical model that is parsimonious over a model that is more complex. A second reason is that with an increasing number of attributes, the chances increase that the combination of levels leads to unrealistic cases. In the next section we will further discuss this issue.

Realism of the Scenarios

Although it is assumed that choice to use a technology or service is based on rational argument, research has shown that in practice decision making is less straightforward. Simon already introduced the concept of bounded rationality (Simon, 1957). Cohen, March, and Olsen (1972) see decision making as a kind of garbage can. Miller, Hickson, and Wilson (1996) stress the political nature of decisions, while others see decision making more as a chaotic process. Although these visions are more related to formal organizational decision-making processes, it is clear that also in day-to-day practices decision making is not that straightforward, as is also illustrated by the contrast between objective and more subjective approaches of media choices. Although we don't focus on adoption theory and media choice models, per se, we want to make clear in which conceptual tradition we want to place our research.

From decision-making theory it is known that individuals, due to cognitive limitations, only use a relatively limited number of criteria to base their decisions upon. Hence, it is important to identify and select relevant criteria to be included in the scenarios. Hypothetical scenarios should include salient variables and combinations that are representative of those observed in reality. Realism can be enhanced by using actual scenarios or by involving knowledgeable individuals and by selecting variables that are naturally orthogonal. Intercorrelation between explanatory variables should be minimized. It is important to use multiple approaches—focus groups, interviews, and participatory observation—to select the proper criteria to be included in the design of the scenarios.

Nevertheless, using multiple salient criteria might result in scenarios that are not realistic in nature. Pre-testing scenarios in which experts as well as potential respondents review the scenarios in terms of their realistic nature, both with regard to the dimensions or attributes used and the combinations of dimensions and attributes, as well as the realistic nature of their levels, is important. Although the use of hypothetical scenarios might be relevant seen from a theoretical point of view, we opt for "realistic" scenarios. Furthermore in the presentation of the scenarios one has to be careful in the semantic representation of the scenarios. In many research projects the cases are clear-cut. If price is a case, dimension stating that the price of a product is €100 is sufficient to create a realistic story. However, in some research projects the value of a dimension is not so explicit. In order to test to what degree the created cases represent the situation it is supposed to, researchers can provide a small number of test respondents with the cases. They are asked to state how a case scores on each of the dimensions in the case. In the case description in the next section, an example of this approach will be presented. Based on the answers of these test respondents the cases can be formulated more or less extreme.

Measuring the Judgment

An important part in the design of a policy capturing study is the way in which the judgment (the dependent variable) is measured. There are basically three ways of measuring the judgment:

- **Choosing strategy**: A choice between several alternatives. This strategy may lead to a dependent variable measured on the nominal level.
- **Ranking strategy**: Compare an alternative relative to other alternatives. This strategy leads to a dependent variable measured on the ordinal level.
- **Rating strategy**: Rate the degree to which an alternative is applicable, suitable to the described situation. This strategy leads to a dependent variable measured on the ratio or interval level.

The choice for either one of these strategies has advantages and disadvantages. Of course the choice of how to measure the judgment strongly depends on the research question. For example, a research question that focuses on in what situation people are likely to adopt a specific technology would require the rating strategy. If, however, the research question focuses on understanding behavior of people in a certain situation, the choosing strategy might be a more logical alternative. An example of the research question could be: "Which PDA is most suited for a certain person in a certain situation." A more sophisticated method is to rank these alternatives (using complete ranking or pair-wise comparison). Choosing one alternative is now compared with choosing one or more other alternatives. For example, PDA X is the least suited for this scenario and PDA Z is the most suited. This, however, could become a tedious job for respondents if a lot of alternatives are involved.

One of the most important consequences of using either the rating, ranking, or choosing strategy is that it influences the possibilities for statistical analysis. With the choosing strategy researchers can use only measures on the nominal level of analysis. If the researcher wants to build an explanatory model he needs to use logistic regression as a method of analysis (see section on data analysis for further explanation of this method). When the researchers choose the rating strategy, more opportunities, such as Ordinary Least Square regression, for statistical data analysis are open. Researchers should bear in mind, however, that a scale rating (like ranking) is only a measurement on the ordinal level. In order to overcome this problem researchers could consider developing a scale. Of course, in that case, the researchers should develop and pretest a suitable scale in order to measure the judgment in a reliable way (using standard scaling techniques such as Likert scales and Cronbachs' Alpha). An alternative and simple solution is to ask respondents to assess the likelihood with which they are going to make use of a specific technology, for instance, in terms of the percentage in which they are likely to use a specific device (Bouwman, 2004).

The downside of ranking and rating multiple alternatives, however, is that the researcher puts a strong claim on the effort of the respondents. Sometimes respondents are willing to put a lot of energy in providing the researcher with information, but most of the time respondents prefer questionnaires that can be filled out quickly. A way to overcome this problem could be to ask respondents which of the alternatives they would consider (yes or no) and what the likelihood is that an alternative would eventually be chosen.

Karren et al. (2002) suggest that scenarios that contain many factors are cognitively complex and may require some practice with the response scale. A two-step approach in which initially a more limited number of cues can be offered, before the full presentation of all the scenarios takes place might be advantageous. This also offers the opportunity to do a limited test-retest check in order to estimate the reliability of the respondents' judgments.

Number and Presentation of Scenarios

As discussed, a major problem is the number of scenarios that a respondent has to respond to. In policy capturing studies it is assumed that respondents read all the scenarios. In itself this is a time-consuming activity that leads to overload, fatigue, loss of interest, and non-response. If the respondents have to deal with more than 30 alternative scenarios, respondents will be wearied out, according to Gustafsson et al. (2003). Although there doesn't appear to be a golden rule, the literature suggests that the stress level with eight scenarios is still low, but, even with more scenarios, stress level remains relatively moderate (3.28 on a 7-point scale, Graham et al., 2001). The range of scenarios that can be offered to a single respondent is somewhere between eight and 64. Of course, it also depends on the length and complexity of the scenarios that are presented.

There are basically two ways the stimuli are presented verbally and visually. But even when one format is selected, the way the scenarios are represented can vary. For example, in some studies the scenarios are presented as small stories or storyboards, while in other projects only key words, descriptive sentences, or combinations of both are being used. In the case of visual representation a distinction has to be made between the real thing and a photo or drawing (graphical representation) of the real thing. Research in the way scenarios are presented in relation to validity of conclusions is not available.

Data Collection Procedure and Sample Type and Size

Directly related to the way the scenarios are presented to the respondent is the question of how the data are collected. If visual representations are used, it is clear that the way the data are collected has to be attuned. Basically, there are several alternatives: face-to-face interview, mail, computer, possibly in combination with the use of Internet, or a telephone. Basically, here normal rules for the use of data collection apply. From a practical point of view, computer-aided support for generating the scenarios makes the use of computer-based interviews via the Internet very attractive. In the discussion of sample size we have to keep in mind that there are two perspectives: the sample of scenarios and the sample of respondents. To increase the power of the results it is important to increase the number of respondents in the study. Basically, also here the same rules as in a regular research project apply.

Data Analysis

An important issue to bear in mind when analyzing the data of a FS/PC/CM study is the goal of the research. The goal can either be of a more practical nature, seeking answers with regard to, for example, the diffusion of a specific technology or the goal could be of a more scientific nature, seeking an explanation for the choices that people make. If the goal is of a more practical nature and we want to predict how often certain choices are made, it is important to remember that the factorial object universe is not a representative sample of situations in daily life. Instead, the factorial object universe represents all possible situations, each of them occurring only once. A consequence of this: it is not possible to generalize to a larger population directly from the results. For example, if the results of the research show that PDA X was chosen in 25% of all scenarios, that does not mean that in real life this PDA will be chosen an equal number of times, because in daily life, certain combinations will occur much more frequently than others. A possible solution to this problem is to ask respondents how likely it is they will be in the situation described by the scenario. Using this rating as a weight factor, careful generalizations can be made. When the goal of the research is to understand what explains certain judgments, we need to find out what the explaining force is of the different independent variables. Depending on the way the data was gathered, there are different methods of data analysis (Table 1). Judgments and evaluations are based on individual and social determined characteristics. Consequently, in the analysis basically two types of predictor variables can be used: situational characteristics and/or individual characteristics. In, for example, an OLS regression, this can be generalized to [2]. Researchers should take possible interaction effects between attributes to be evaluated and judge characteristics into account.

$$J_i = b_0 + \sum b_k X_{ik} + \ldots + \sum b_r C_{ir} + \varepsilon_i \qquad \text{[Equation 2]}$$

Table 1. Data Analysis.

		Number of vignettes per respondent	
		One	More than one
Dependent variable	Nominal	Logistic regression	Multilevel logistic regression
	Ratio or interval	Regression analysis	Hierarchical linear modeling

Conclusion

With regard to the results of this theoretical and methodological approach, we would like to refer to a number of articles that have been published over the past 10 years (an overview is provided in Appendix I). We can conclude that the approach taken in this research project allowed us to gain insight into the circumstances in which people with different characteristics would like to use this specific technology.

We have described a research method that is well known in a number of research areas, but is little used in the area of adoption and use of new information and communication products and services. In this research method, based on factorial surveys, policy capturing, and vignette studies, the basic idea is to present people with fictional cases that they respond to. By systematically varying the cases, we can see the effect on the degree to which people may like to use an IT product or service. Essentially, because people are not familiar with the new service, but are familiar with real life scenarios, the method provides an ideal starting point for understanding the potential success of IT services and products that do not exist on the market on a large scale. Although a number of design issues have to be dealt with using the method, we believe that the results provide reliable and valid results.

The implications of this way of doing research with regard to theory are that a focus on TAM-concepts, such as ease of use, perceived usefulness, or Diffusion of Innovation concepts, may be too generic to help us understand the mechanism at play in understanding the use of new technologies. Context-related variables are more likely than personal characteristics to define usage behavior. This research method provides the opportunity to gain insight into the underlying factors of perceived usefulness. From a product design point of view, using an approach based on vignette type of studies or conjoint analysis helps us better understand the potential use of new technologies and allows us to pre-test concepts of new technologies and their potential usefulness in more detail in an early stage of diffusion.

Appendix I: Overview of Situational Approach Papers

Baaren, E., Van de Wijngaert, L. & Huizer, E. (2009). *Not another TAM paper: Relating individual and context characteristics to the adoption of HDTV*. Paper presented at the 59th Annual Conference of the International Communication Association.

Bouwman, H., Haaker, T., & De Vos, H. (2008). *Mobile applications for police officers*. Paper presented at the 21st Bled eConference: Overcoming Boundaries through Multi-Channel Interaction, June 15–18, Bled, Slovenia.

Bouwman, H., & Van de Wijngaert, L. (2009). Coppers, context, and conjoints: A reassessment of TAM. *Journal of Information Technology, 24*(1), 1–19.

Bouwman, H., Van de Wijngaert, L., & De Vos, H. (2008). *Context-aware mobile services for police officers*. Paper to be presented at the International. Paper presented at the Conference on M-business.

Haaker, T., de Vos, H., & Bouwman, H. (2006, June 5–7). *Mobile service bundles: The example of navigation services*. Paper presented at the 19th Bled Electronic Commerce Conference on eValues, Bled, Slovenia.

Haaker, T., de Vos, H., & Bouwman, H. (2007). Mobile service bundles: The example of navigation services. *Electronic Markets, 17*(1), 28–38.

John, R., Simons, L., & Bouwman, H. (2009, June 14–17). Designing and testing service experiences (mobile, web, public displays) for airport transit. Paper presented at the 22nd Bled eConference eEnablement: Facilitating an Open, Effective and Representative eSociety, Bled, Slovenia.

Sperling, R., Simons, L., & Bouwman, H. (2007). *Multi-channel service concept design and prototyping*. Paper presented at the McMaster Conference.

Steenbeek, W., Van de Wijngaert, L., Van den Brand, M., Brinkkemper, S., & Harmsen, F. (2005). *Sourcing decision-making: Elicitating consultancy knowledge using policy capturing*. Paper presented at the 13th European Conference on Information Systems: Information Systems in a Rapidly Changing Economy, Regensburg.

Walther, J. B. (1992). Interpersonal effects in computer-mediated interaction: A relational perspective. *Communication Research, 19*(1), 52-90.

Van de Wijngaert, L. (1998). *A policy capturing study of media choice*. Paper presented at the Conference Information Seeking in Context, Sheffield, UK.

Van de Wijngaert, L. (1999). *Matching Media: Information need and new media choice*. Enschede: Telematica Institute.

Van de Wijngaert, L., & Bouwman, H. (2002). Content and context: A new research approach to the basic characteristics of information needs. *New Media and Society, 4*(3), 329–354.

Van de Wijngaert, L., & Bouwman, H. (2008). Would you share? Predicting the potential use of a new technology. *Telematics and Informatics, 26*(1), 85–102.

Vermaas, K., & Van de Wijngaert, L. (2003). The disclosure of personal information on the Internet: User motivation, reliability and price as explaining factors. In O. Petrovic, M. Ksela, M. Fallenbock & C. Kittl (Eds.), *Trust in the network economy* (Vol. 2, pp. 365–374). Berlin: Springer-Verlag.

References

Ajzen, I. (1991). The theory of planned behavior. *Organizational Behavior and Human Decision Processes, 50*(2), 179-211.

Amberg, M., Hirschmeier, M., & Wehrmann, J. (2004). The compass acceptance model for the analysis and evaluation of mobile services. *International Journal Mobile Communications*, *2*(3), 248–259.

Benbasat, I., & Barki, H. (2007). Quo Vadis TAM? *Journal of the AIS*, 8(3), 211–218.

Bijker, W. E. (1997). *Of bicycles, bakelites, and bulbs: Toward a theory of sociotechnical change*. Boston, MA: MIT Press.

Blechar, J., Knutsen, L., & Damsgaard, J. (2005). *Reflexivity, the social actor and m-service domestication: Linking the human, technological and contextual*. Paper presented at the Designing Ubiquitous Information Environments: Socio-Technical Issues and Challenges, Cleveland, Ohio.

Bouwman, H. (2004). Anywhere, anytime, anyplace or here and now in this context: 3rd generation Mobile Services: a Vignet-study. Paper presented to 54th annual Conference of the ICA.

Bryk, A. S., & Raudenbush, S. W. (1992). *Hierarchical linear models: Applications and data analysis methods*. Newbury Park, CA: Sage.

Carlson, J. R., & Zmud, R. (1997). Channel expansion theory and the experiential nature of media richness perceptions. *Academy of Management Journal*, *42*(2), 153–170.

Chan-Olmstead, S. M., & Chang, B. (2006). Audience knowledge, perceptions and factors affecting the adoption intent of terrestrial digital television. *New Media & Society*, 8(5), 773–800.

Ciborra, C. & Willcocks, L. (2006). The mind or the heart? It depends on the (definition of) situation. *Journal of Information Technology*, 21 (3), 129-139.

Cohen, M. D., March, J. I. G., & Olsen, J. P. (1972). A garbage can model of organizational choice. *Administrative Science Quarterly*, *17*, 1–25.

Cyr, D., Head, M., & Ivanov, A. (2006). Design aesthetics leading to m-loyalty in mobile commerce. *Information & Management* 43, 950–963.

Daft, R., & Lengel, R. (1984). Information richness: A new approach to managerial behavior and organizational design. In L. Cummings & B. Staw (Eds.), *Research in Organizational Behavior* (Vol. 6, pp. 191–133). Greenwich, CT: JAI.

Daft, R., & Lengel, R. (1986). Organizational information requirements, media richness and structural design. *Management Science*, *32*(5), 554–571.

Davis, F. D. (1986). *A technology acceptance model for empirically testing new end-user information systems: Theory and results*. Massachusetts: Sloan School of Management, Massachusetts Institute of Technology.

Davis, F. D. (1989). Perceived usefulness, perceived ease of use and user acceptance of information technology. *MIS Quarterly*, *13*(3), 319–340.

DeSanctis, G., & Poole, M. (1994). Capturing the complexity in advanced technology use: Adaptive structuration theory. *Organization Science*, 5(2), 121–147.

Dishawa, M. T., & Strong, D. M. (1999). Extending the technology acceptance model with task–technology fit constructs. *Information & Management*, 36(1), 9–21.

Fishbein, M., & Ajzen, I. (1975). *Beliefs, attitude, intention and behavior: An introduction to theory and research*. Reading, MA: Addison-Wesley.

Frissen, V. (2000). ICTs in the rush hours of life. *Information Society*, *16*(1), 65–75.

Fulk, J., Schmitz, J., & Steinfield, C. (1990). A social influence model of technology use. In J. Fulk & C. Steinfield (Eds.), *Organizations and Communication Technology* (pp. 117–142). Newbury Park, CA: Sage.

Goodhue, D. L., & Thompson, R. L. (1995). Task-technology fit and individual performance. *MIS Quarterly*, *19*(2), 213–236.

Graham, M., & Cable, D. (2001). Consideration of the incomplete block design for policy capturing research. *Organizational Research Methods*, *4*(1), 26–45.

Gustafsson, A., Herrmann, A., & Huber, F. (2003). Conjoint analysis as an instrument of market research practice. In A. Gustafsson, A. Herrmann & F. Huber (Eds.), *Conjoint measurement. Methods and applications*. Berlin: Springer-Verlag.

Heres, J., Mante-Meijer, E., Turk, T., & Pierson, J. (2005). Adoption of ICTs: a proposed framework. In E. Mante-Meijer & L. Klamer (Eds.), *Capabilities in action: What people do*. Luxembourg: Office for Official Publications of the European Communities.

Hijink, M. (2008, 30 May). Tv-kijker nu wel HD-ready [TV Viewer HD-Ready]. *NRC Handelsblad*. Retrieved from http://www.nrc.nl/media/article1903584.ece/Tv-kijker_is_nu_wel_HD-ready

Hirschman, E. C. (1980). Innovativeness, novelty seeking, and consumer creativity. *Journal of Consumer Research*, *7*, 283–295.

Ilie, V., Van Slyke, C., Green, G., & Lou, H. (2005). Gender differences in perceptions and use of communication technologies: A diffusion of innovation approach. *Information Processing Management Journal*, *18*(3), 13–31.

June, L., E., Y. J., & Yu, C. (2005). Personal innovativeness, social influences and adoption of wireless Internet services via mobile technology. *Journal of Strategic Information Systems*, *14*(3), 245–268.

Karren, R., & Woodard Barringer, M. (2002). A review and analysis of the policy-capturing Methodology in Organizational research: Guidelines for Research and Practice, *Organizational Research Methods* 5(4), 337–336.

Kleijnen, M., De Ruyter, K., & Wetzels, M. (2004). Consumer Adoption of wireless services: Discovering the rules while playing the game. *Journal of Interactive Marketing*, *18*(2), 51–61.

Legris, P., Ingham, J., & Collerette, P. (2003). Why do people use information technology? A critical review of the technology acceptance model. *Information & Management*, 40(3), 191–204.

Lievrouw, L., & Livingstone, S. (2002). Introduction, the social shaping and consequences of ICTs. In L. Lievrouw & S. Livingstone (Eds.), *The handbook of new media* (pp. 1–15). London: Sage.

Lu, J., Yu, C.S., Liu, C., & Yao, J. (2003). Technology acceptance model for wireless Internet. *Internet Research: Electronic Networking Applications and Policy*, *13*(3), 206–222.

Manning, K. C., Bearden, W. O., & Madden, T. J. (1995). Consumer innovativeness and the adoption process. *Journal of Consumer Psychology* 4(4), 329–346.

Mante-Meijer, E., & Haddon, L. (2001). *Checking it out with the people. ICT markets and users in Europe*. Heidelberg: Eurescom.

Mao, E., Srite, M., Thatcher, J., & Yaprak, O. (2005). A research model for mobile phone service behaviors: Empirical validation in the U.S. and Turkey. *Journal of Global Information Technology Management*, 8(4), 7–28.

Markus, L. (1990). Towards a "critical mass" theory of interactive media. In J. Fulk & C. Steinfield (Eds.), *Organizations and communication technology* (pp. 94–118). New York: Sage.

Martocchio, J. J., Webster, J., & Baker, C. R. (1993). Decision-making in management information systems research: The utility of policy capturing methodology. *Behavior & Information Technology, 12*(4), 238–248.

McCreadie, M., & Rice, R. E. (1999a). Trends in analyzing access to information. Part I: Cross-disciplinary conceptualizations of access. *Information Processing and Management, 35*(1), 45–76.

McCreadie, M., & Rice, R. E. (1999b). Trends in analyzing access to information. Part II. Unique and integrating conceptualizations. *Information Processing and Management 35*(1), 77–99.

Miller, S. J., Hickson, D. J., & Wilson, D. C. (1996). Decision making in organizations. In S. R. Clegg, C. Hardy & W. R. Nord (Eds.), *Handbook of organization studies* (1 ed., pp. 293–312). London: Sage.

Molin, E. (1999). *Conjoint modeling approaches for residential group preferences*. Eindhoven: Eindhoven University of Technology.

Monge, P. R., & Contractor, N. S. (2003). *Theories of communication networks*. New York: Oxford University Press.

Olson, M. (1965). *The logic of collective action. Public goods and the theory of the group*. New York: Schocken.

Poole, M., & DeSanctis, G. (1990). Understanding the use of group decision support systems: The theory of adaptive structuration. In J. Fulk & C. Steinfield (Eds.), *Organizations and communication technology*. (pp. 173–193). Newbury Park, CA: Sage.

Postma, J., & Wallage, J. (2007). Het uur van de warhead [The hour of truth]. Advise of the committee Postma/Wallage on the direction of the eGovernment. The Hague: e-overheid.

Rogers, E. M. (2003). *Diffusion of innovation* (5 ed.). New York: The Free Press.

Rossi, P. H., & Anderson, A. B. (1982). The factorial survey approach: An introduction. In P. H. Rossi & S. L. Nock (Eds.), *Measuring social judgments* (pp. 15–67). Beverly Hills: Sage.

Rossi, P. H., & Nock, L. (Eds.). (1982). *Measuring social judgments*. Beverly Hills, CA: Sage.

Schepers, J., & Wetzels, M. (2007). A meta-analysis of the technology acceptance model: Investigating subjective norm and moderating effects. *Information & Management, 44*, 90–103.

Schmitz, J., & Fulk, J. (1991). Organizational colleagues, media richness, and electronic mail: A test of the social influence model of technology use. *Communication Research, 18*(4), 487–523.

Silverstone, R., & Haddon, L. (1996). Design and the domestication of information and communication technologies: Technical change and everyday life. In R. Mansell & R. Silverstone (Eds.), *Communication by design. The politics of information and communication technology*. Oxford: Oxford University Press.

Simon, H. (1957). *Models of man*. New York: Free Press.

Straub, D. W., & Burton-Jones, A. (2007). Veni, Vidi, Vici: Breaking the TAM logjam. *Journal of the AIS*, 8(4), 223–229.

Van der Heijden, P. (2003). User acceptance of hedonic information systems. *MIS Quarterly, 28*(4), 695–704.

Venkatesh, V., & Davis, F. D. (2000). A theoretical extension of the technology acceptance model: Four longitudinal field studies. *Management Science, 46*(2), 186–204.

Venkatesh, V., Morris, M. G., Davis, G. B., & Davis, F. D. (2003). User acceptance of information technology: Toward a unified view. *MIS Quarterly, 27*(3), 425–478.

Vermaas, K., & Van de Wijngaert, L. (2005). Measuring Internet behavior: Total time diary and activity diary as research methods. *Journal of Information Technology Theory and Application, 7*(1), 121–134.

Vriens, M. (1995). *Conjoint analysis in marketing. Developments in stimulus representation and segmentation methods*. Groningen: Groningen University.

Vriens, M. (1998). Verbal versus realistic pictorial representation in conjoint analysis with design attributes. *Journal of Product Innovation Management, 15*, 455–467.

Webster, J., & Trevino, L. K. (1995). Rational and social theories as complementary explanations of communication media choices: Two policy-capturing studies. *Academy of Management Journal, 38*(6), 1544–1572.

Williams, R., & Edge, D. (1996). The social shaping of technology. *Research Policy, 25*(6), 865-899.

Zedeck, S. (1977). An information processing model and approach to the study of motivation. *Organizational Behavior and Human Performance, 18*, 47–77.

· 5 ·

Mathematical Models of the Diffusion Process

GEORGE A. BARNETT

Introduction

Beginning with the classical S-shaped diffusion curve, this chapter reviews various mathematical models including the internal, external, and the mixed influence/fundamental models that have been used to describe the process of the diffusion of innovations. This chapter discusses the assumptions on which these models are based. It then suggests alternative models that make different assumptions and which may provide a better description of the diffusion process. These include the r-shaped diffusion curve, the adoption and discontinuance model, media substitution, and generational diffusion models. The chapter concludes by suggesting that future research should employ Fourier analysis to determine innovation life cycles and autoregressive models to examine the impact of displacement or supplemental effects of later generations of innovations.

It is almost an article of faith among communication scholars and other social scientists that the diffusion of information and innovations through society over time resembles an S-shaped sigmoid or a logistic curve. That is, the process begins slowly, accelerates, reaching its maximal rate of knowledge or adoption when half the population has adopted or becomes knowledgeable. At this point, the process begins to decelerate, eventually leveling off when it reaches an upper asymptote (the limits of the population). There are no new individuals to adopt the new product or practice or learn the new idea. Thus, the diffusion process is complete. This conceptualization of the diffusion curve is an oversimplification and is based on a number of assumptions about the adopting population, the channels through which information is disseminated, and characteristics of the innovation. This chapter reviews various mathematical models that have been used to describe the diffusion process and discusses the assumptions on which they are based. It then suggests alternative models that make different assumptions and which may provide better descriptions of the diffusion process.

The S-Shaped Curve

The S-shaped or sigmoid curve is a special case of the logistic function or logistic curve. It is graphically represented in Figure 1. A simple logistic function may be defined by equation 1

$$P(t) = \frac{1}{1 + e^{-t}} \qquad \text{[Equation 1]}$$

where the variable P denotes the members of a population and t represents time. Formula 1 models the S-curve of growth of P, where P is the population of adopters or knowers, or in the case of epidemics, the number of individuals stricken by a disease. The initial stage of growth ($0 < t < ½$) is an exponential, then, because as saturation begins (at 50%), the curve becomes a decaying exponential ($½ < t < 1$), the growth slows, and, at maturity ($t = 1$), growth stops. Thus, formula 1 may be decomposed into two equations of limited range.

$$P(t) = be^{kt} \qquad \text{[Equation 2]}$$

for the initial stage of growth ($0 < t < ½$), and

$$P(t) = be^{-kt} + kn \qquad \text{[Equation 3]}$$

to describe the decline in the growth rate during the second half of the diffusion process ($½ < t < 1$). k may be considered the diffusion coefficient, and n, the number of people who have adopted or those who know the new information. An exponential curve and a decaying exponential are graphically represented as Figures 2 and 3. In the model expressed as equation 1 there is an inflection point when 50% of the population adopt the innovation. At this point, the maximum rate of adoption occurs. There are alternative models of diffusion in which the inflection point is flexible, i.e., the maximum rate of diffusion does not occur when 50% of the population adopts (Mahajan &

Peterson, 1985).

Since any unimodal distribution function will generate an S-shaped curve, it

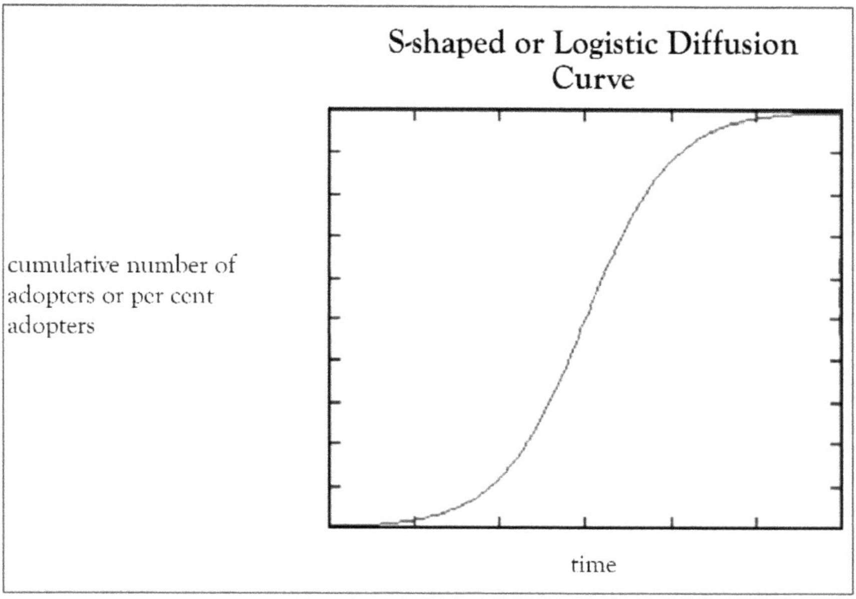

Figure 1

is often not possible to empirically determine which of several competing functions best describes a given diffusion curve. Therefore, attempts have been made to develop theory-based models for analyzing and modeling the diffusion process (Bass, 1969; Mahajan & Peterson, 1985).

Dodd (1955) presented a mathematical model that used a logistic function to describe diffusion. It begins with the differential equation

$$dp/dt = kpq \qquad \text{[Equation 4]}$$

where p is the proportion of knowers or adopters in the population, $q = 1\text{-}p$ (non-adopters), and k is a constant or the probability that an interaction between a knower and non-knower will result in the non-knower becoming a knower. To determine the accumulated number of knowers (or adopters) at any point in time, equation 4 may be integrated to produce

$$P_t = \frac{1}{1 + (q/p)e^{kt}} \qquad \text{[Equation 5]}$$

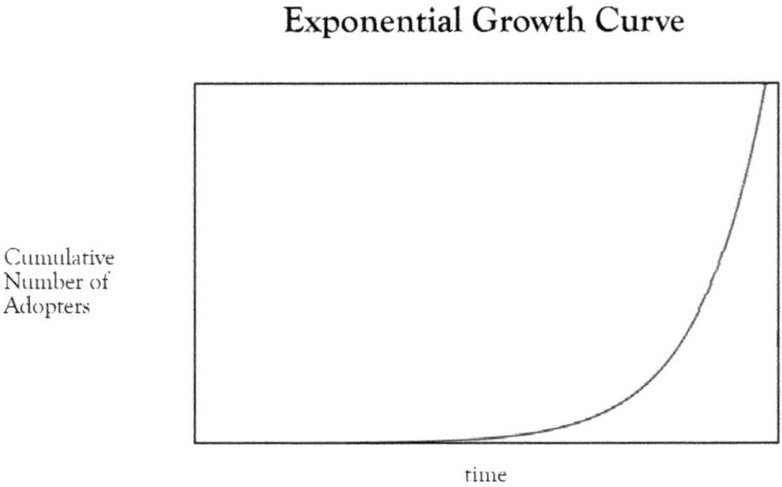

Figure 2

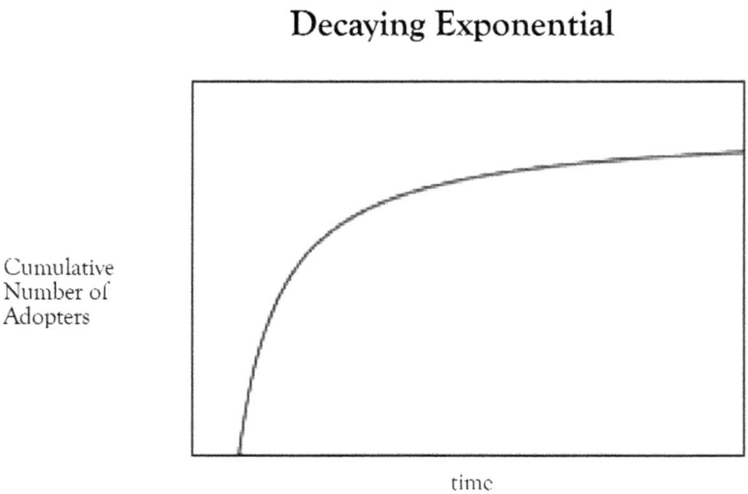

Figure 3

The Internal-Influence Model

Coleman (1964) proposed a model in which the rate of diffusion is proportional to the number of people who have already adopted the innovation. People who have already adopted persuade others to adopt at a constant rate. Alternatively, individuals may choose to adopt based on their observations of those who have already implemented the innovation. This may be simple imitation or through social learning (Bandura, 1977). It is described by the following differential equation:

$$dA/dt = kA_t\,(N\text{–}A_t) \qquad \text{[Equation 6]}$$

where A is the number of adopters, *N*, the population size, and *k*, a constant rate of change or coefficient of diffusion. Thus, the rate of diffusion at any point in time is a function of the difference between the total number of potential adopters at that time and the number of previous adopters (*N*–A). The specific value of *k* depends on the various aspects of the diffusion process, including attributes of the innovation, the channels of communication employed, and characteristics of the social system (Rogers, 2003). This model assumes that the population is homogenous on those characteristics that have an impact on the diffusion process. Rogers (2003) discusses in great detail the elements of diffusion that determine the value of *k*. Integrated, the predictive equation becomes

$$A_t = \frac{Ne^{kt}}{N\text{–}1 + e^{kt}} \qquad \text{[Equation 7]}$$

Because this model proposes that the rate of diffusion is a function of the disparity between the number of potential adopters and the number of previous adopters, it results in an S-shaped curve (Jackson, 2008). The initial adopters (innovators) are those who spontaneous adopt. There is no one to persuade them or for them to copy. Thus, the early rate of adoption is very slow. As the process develops, there are more individuals to learn from, which leads to an increase in the rate of diffusion. Over time, since more have adopted, there are fewer to be persuaded and the process slows, until eventually it becomes asymptotic with the number of potential adopters simply because there are no individuals who have not adopted and the process ends.

This model has come to be known as the internal-influence model (Mahajan & Peterson, 1985) because it suggests that diffusion occurs only through interpersonal communication (persuasion or imitation) between prior adopters and potential adopters ($(A_t)(N\text{–}A_t)$). In network terms, this is a contagion model of diffusion (Burt, 1987). Thus, k represents the effect of horizontal, decentralized, or informal channels of communication (Mahajan & Peterson, 1985).

The External-Influence Model/ The R-Shaped Curve

Coleman (1964) proposed a second model to describe the diffusion process. When diffusion occurs within a finite population and information emanates from a constant source, such as the mass media independent of the number of people who already have adopted, the number of adopters per unit time is proportional to the total number who has not yet adopted the innovation. The accumulation of adoptions will increase as a decaying exponential of time. This process may be described by the differential equation

$$dA/dt = k\,(N - A_t) \quad \text{[Equation 8]}$$

In this case, k is the coefficient of external influence. At t equal to 0 and A equal to 0, the predictive equation becomes

$$A_t = be^{-kt} \quad \text{[Equation 9]}$$

Adding the initial adopters or knowers, equation 9 becomes

$$A_t = be^{-kt} + kn \quad \text{[Equation 10]}$$

the same as the decaying exponential (equation 3) that describes the later portion of the diffusion process.

This model has become known as the external-influence model (Mahajan & Peterson, 1985) since the constant term emanates from outside the social system. The diffusion constant typically represents the influence of the mass media, the government, change agents, or a salesperson. Abrahamson and Rosenkopf (1996) point out that external influence theories also have been found to explain the diffusion of administrative innovations when government agencies are actively involved in mandating the use of these techniques. It may be considered a formal channel of communication from a central source (Mahajan & Peterson, 1985). The model assumes that the rate of diffusion is dependent only on the number of potential adopters. Unlike the internal-influence model it does not attribute diffusion to the interaction between those who have already adopted and those who have not. The external-influence model is best applied when the members of a social system are physically isolated. As a result, there is little social interaction (Coleman, Katz & Menzel, 1966). Their social networks are very sparse and play a minor role in the diffusion process (Valente, 1994; Abrahamson & Rosenkopf, 1996). The external-influence model is suitable when the innovation is not complex or when information about the innovation is available only from sources external to the social system (Hamblin et al., 1973).

The external-influence model produces a curve that resembles an "r" rather than the S-shaped curve typically used to describe the diffusion process. The external agent forces adoption nearly simultaneously. As a result, the adoption curve begins convexly with the maximum growth rate slowing only when most potential adopters have adopted, thus resembling an r. External-influences have been found in a number of diffusion studies (Rogers, 2003). For example, Hamblin et al. (1973) reported that it was appropriate to describe the diffusion of labor strikes and political assassinations among geographically separated countries. Rosengren (1987), in a cross-cultural study of news diffusion, found that the various media systems played important roles in the rate of dissemination and that the processes did not resemble an S-shaped curve. Rather, due to the rapid diffusion via the mass media, the rate of learning appeared to be r-shaped.

Danowski, Gluesing, and Riopelle (2009) found support for the hypothesis that

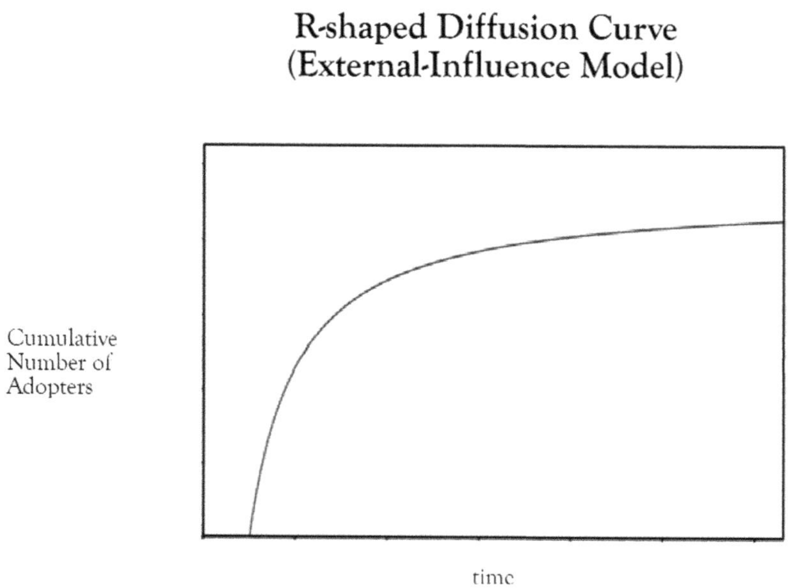

Figure 4

innovations diffused through heavy use of ICTs have a convex r-curve rather than the traditional S-curve. They suggest the following reasons for this pattern of adoption: (1) There is heavy ICT use in the dissemination of information about the innovation; (2) visibility of the number of adopters and their status in the adopting organization; (3) ambiguity about the innovation's attributes; (4) herd effects, a contagious emotional and cognitive phenomenon that results in nearly everyone thinking, deciding, and acting in the same way; (5) information cascades, when people observe the actions of others and then make the same choice that the others have made, independently of their own private information signals; (6) reduced interpersonal communication (a necessary condition for logistic diffusion); and (7) in some cases, mandated adoption.

This represents a major shift in how innovations may diffuse within social systems that have well-developed ICT-based communication environments. The ICT acts in the same manner as an external source. The adoption process is more rapid without much direct interpersonal communication than in traditional systems characterized by the slowly building S-shaped curve that requires potential adopters to directly communicate with earlier adopters and deciding whether or not to adopt. Perhaps the real significance of the external-influence model and Danowski et al.'s research is that there are numerous theoretical alternatives to the S-shaped curve to describe adoption (or the dissemination of information) over time.

The Mixed-Influence or Fundamental Diffusion Model

Both the internal-influence and the external-influence models represent idealized cases. More realistically, the diffusion of innovation and the dissemination of information take place through external mass media and interpersonal communication channels within the social system. As Rogers (2003) points out diffusion typically involves the mass media and interpersonal communication with the media primarily informing potential adopters about the innovation and interpersonal networks being more important in persuading individuals to adopt or reject the innovation. Thus, the mixed-influence model that subsumes both the previous models,

$$dA/dt = (a + k\,A_t =)(N - A_t), \qquad \text{[Equation 11]}$$

where a is equivalent to k in the external-influence model. Integrated, the mixed influence model becomes,

$$A_t = \frac{N-[a\,(N-A_0)/(a + k\,A_0\,)]\,\exp[-(a + k\,A_t)(t-t_0)]}{1 + [k(N-A_0)/(a + k\,A_0)]\,\exp[-(a + k\,A_t)(t-t_0)]}. \quad \text{[Equation 12]}$$

The plot of the cumulative adopter distribution results in a generalized logistic (S-shape) curve. The coefficient of external-influence a and the internal-influence index k determine the shape of the curve. The mixed-influence model is also known as the fundamental diffusion model (Mahajan & Peterson, 1985) or the Bass model (Bass, 1969), which has been applied to predict the marketing of new products (Mahajan, Muller & Bass, 1990).

Assumptions of the Model

There are several assumptions underlying the set of models described above. Perhaps the most important is that the diffusion process is binary. Members of a social system either adopt an innovation (know the new information) or not. One cannot adopt (or know) just a little bit. This is a discrete rather than a continuous process (Mahajan & Peterson, 1985). As such, it is not appropriate for the investigation of cultural or attitudinal processes (Barnett, 1988) or for the study of the stages in the diffusion process (Rogers, 2003).

The model considers only a single adoption by the members of the social system (Mahajan & Peterson, 1985). Individuals cannot adopt multiple times. They cannot discontinue its use and/or readopt. They cannot learn some new information, forget that knowledge, and relearn it, or in the epidemiological case, one cannot contract the same disease twice. Further, these models make no prevision for the discontinuance of an innovation once it has been adopted. As will be shown in the next section of this chapter, Barnett, Fink, and colleagues have developed an alternative model to accommodate this necessary modification (Barnett, Fink & Debus, 1989; Chang & Barnett, 1991; Barnett & Fink, 2008).

Nor does the model allow adopters to modify (reinvent) the innovation or change the information as it is diffused through their social networks. The fundamental diffusion model assumes that an identical innovation is adopted or that information remains an exact copy as it is disseminated. Diffusion scholars recognize that innovations are reinvented, changed, or modified by the user in the process of adoption or implementation. Rogers (2003) provides numerous examples of reinvention with educational, policy, and agricultural innovations. Regarding information dissemination it is well known that as information is exchanged through social networks, it is distorted through the processes of leveling, sharpening, and assimilation (Allport & Postman, 1947).

Further, the innovation is assumed to be independent of other innovations (Mahajan & Peterson, 1985). However, innovations often are comprised of interdependent bundles or collections of components, each of which may be considered a separate innovation. Examples of technology clusters include the Green Revolution composed of new varieties of rice or wheat seed, chemical fertilizers, pesticides, and cultivation practices, or the personal computer with separate innovations in microprocessors, video displays, laser optics for information distribution, and printing (Rogers, 2003). These are innovation clusters (Barnett & Siegel, 1988; Beninson-Germano, Barnett, Dinkelacker & Downing, 1988), and the adoption of one member of the cluster has an impact on the adoption of the others (Vishwanath & Chen, 2006) by either necessitating or limiting adoption of each individual innovation. Thus, this assumption is problematic.

Another assumption underlying the fundamental diffusion model is that there is a known distinct and fixed population of potential adopters in the social system. The size of the social system, N, is constant. It cannot grow or shrink during the course of the diffusion process.

Also, the internal-influence and mixed-influence models assume there is complete mixing among the members of the social system. That is, the likelihood of any two individuals interacting is equally probable. The population is homogenous regarding those attributes that have an impact on the diffusion process. The system's social networks are not differentiated into sub-groups or clusters (Coleman et al., 1966; Rogers & Kincaid, 1981). In other words, there is no social structure—obviously, a fallacious assumption.

Additionally, these models assume that message effects, either due to social interaction or via external sources such as the mass media, are constant regardless of time of adoption or stage in the diffusion process (Mahajan & Peterson, 1985).

Adoption and Discontinuation (Learning and Forgetting)

The fundamental model of diffusion suffers from what Rogers (2003) labels the pro-innovation bias. This is the implication that an innovation should be diffused and adopted by all members of society, that it should be diffused more rapidly, and that it should be neither re-invented nor rejected or its use discontinued. There are two major reasons for discontinuance (Rogers, 2003). The first is dissatisfaction because the innovation is inappropriate for the individual or it does not perform up to expectations. New information may come to light that indicates that the product or practice has serious problems. The innovation may no longer be compatible with changes in the individual or newer innovations. The second is replacement, in which the innovation is rejected in order to adopt one that is perceived as better. The existing product or practice is supplanted by new and improved innovations.

One model that addresses some of the assumptions of the pro-innovation bias was developed by Barnett, Fink, and colleagues (Barnett, Fink & Debus, 1989; Chang & Barnett, 1991; Barnett & Fink, 2008). Barnett, Fink, and Debus (1989) used an exponential growth and decay model to describe the patterns of adoption and discontinuance. The same model may be applied to learning and forgetting.

$$Y_t = a * e^{-(bt)+d} - e^{-ct} \qquad \text{[Equation 13]}$$

where Y_t is the proportion of the population that adopts an innovation or the level of use (or knowledge) at a given time, and *a*, *b*, *c*, and *d* are non-negative constants. Also, *c* is greater than *b*. This model is graphically represented in Figure 5.

Barnett et al. (1989) employed citation data from the sciences, social sciences, and the arts and humanities to examine the robustness of the model. The citation data show a unique pattern of adoption and discontinuance. The initial level of citation starts out small and then increases, reaching a peak in less than two years (1.53 years for the sciences, 1.75 for the social sciences, and 1.16 for the arts and humanities), and decreases over time.

In order to find the trajectory of adoption/discontinuance over time, this model can be differentiated with respect to *c* and setting the derivative as zero. Hence *t**, the time when the theoretical maximum level of adoption occurs, can be obtained.

$$t^* = \frac{ln(c) - ln(b) - d}{c - b} \qquad \text{[Equation 14]}$$

The results of the test of the model on the citation data explained from 97.2% (arts and humanities) to 99.2% (social science) of the variance. Barnett, Fink, and Chang (1988) examined the adoption and discontinuance of motion picture attendance and monochromatic television. Again, the model fit the data well: 98% of the variance for motion picture attendance and 99.9% of the variance for monochromatic television.

This model has been criticized because it predicts that over time the eventual level of adoption will be zero (Kang & Cheng, 1989). The prediction of zero movie attendance did not correspond with historical evidence. To accommodate this criticism, Chang and Barnett (1991) added another parameter, g, to represent the eventual level of adoption.

$$Y_t = a * e^{-(bt)+d_{-e}^{-ct}} + g \qquad \text{[Equation 15]}$$

In those cases where the use of the innovation is eventually totally discontinued, g will equal zero, and there will be no change in the model. The Chang and Barnett revision of the earlier model is represented in Figure 6.

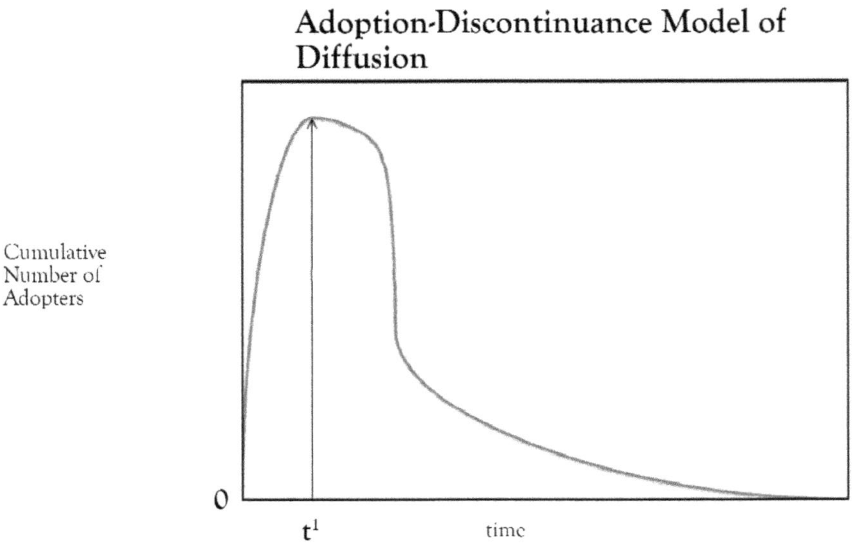

Figure 5

Chang and Barnett (1991) tested the revised model with five different innovations: intercity motor passengers, horses and mules, daily newspaper circulation (in spite of the Internet it still hasn't reached zero), motion picture attendance, and monochromatic television. The proposed model fit the data well, explaining between 95.4% and 99.9% of the variance depending on the innovation. However, the revised model did not represent a significant improvement over the earlier model, except where there is a long flat tail of adoption as in the cases of motion picture attendance and the ownership of horses and mules.

Revision of Adoption-Discontinuance Model

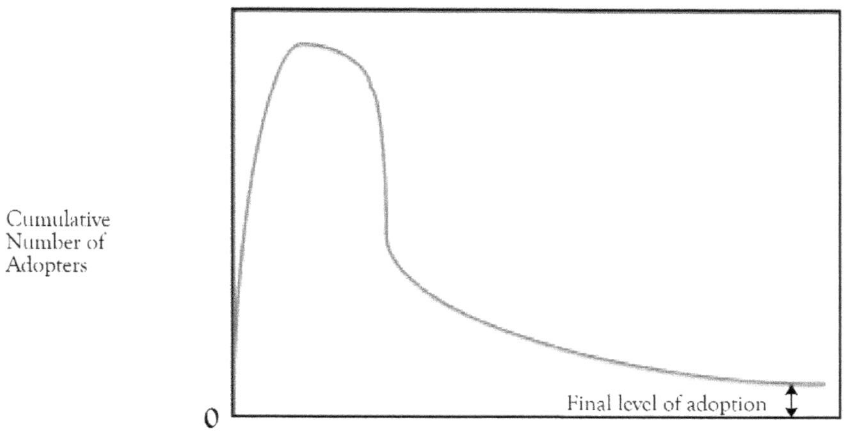

Figure 6

Barnett and Fink (2008) extended the adoption-discontinuance model by building in the ability to include additional variables that have an impact on the rate of diffusion. This was accomplished by adding a parameter, *f*, which alters the rate of adoption and discontinuation and *w* the added variable.

$$Y_t = a * e^{-(bt)+d+fw\_{-e}^{-ct}} \qquad \text{[Equation 16]}$$

This model may be further revised to accommodate any number of additional variables. In equation 17, *u* represents the extra variable, and g, the parameter that determines the rate of diffusion.

$$Y_t = a * e^{-(bt)+d+fw+gu}{}_{-e^{-ct}} \qquad \text{[Equation 17]}$$

Barnett and Fink (2008) evaluated the revised models with sciences and social sciences citation data from 1969 to 2004 to examine the impact of the Internet and scholar age distribution on the citation process. They found that incorporating the *fw* term (equation 16) to represent the affect of the Internet produced a very small improvement in the overall explained variance. For the sciences, it increased from 0.964 to 0.972, and for the social sciences, from 0.981 to 0.982. However, other aspects of the goodness-of-fit improved, such as the skewness and kurtosis of the residuals. When both the Internet and scholar age distribution were added (equation 17), the explained variance rose to 0.996 for the sciences and 0.998 for the social sciences. In all cases, parameters *f* and *g* were statistically significant.

The Barnett and Fink model was developed as an external-influence model. They considered scholarly information to be diffused only through the mass-media, academic journals. The adoption portion of the model may be modified to accommodate innovations diffused through interpersonal interaction by altering the second term (e^{-ct}) to a logistic ($1/1+ e^{-ct}$). Thus, the modified model, expressed in equation 18, would be:

$$Y_t = a * e^{-(bt)+d+fw+gu}{}_{-1/1+e^{-ct}} \qquad \text{[Equation 18]}$$

The model is graphically represented in Figure 7.

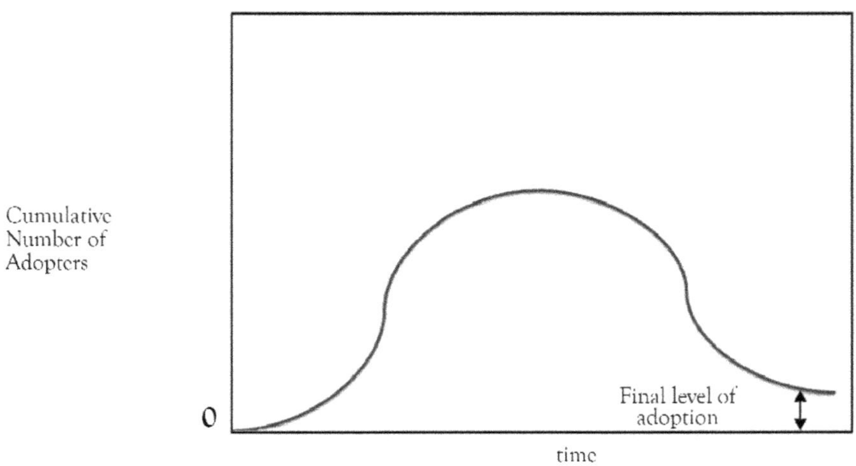

Figure 7: Adoption: Discontinuance Modelwith Logistic.

It is doubtful that equation 18 would provide a better goodness-of-fit than earlier models for the described examples, since there is little area for improvement. However, for innovations diffused primarily through social networks, it may represent a significant improvement.

Media Substitution

There is ample evidence that the use of innovations is discontinued due to dissatisfaction or replacement. Within the area of communication and information technology this process is known as media substitution (Atkin et al., 1998; Jeffres et al., 1995; Lin, 2001). The introduction of a new medium encourages a restructuring of the way people perceive existing media. Media substitution occurs in two ways, through displacement and supplementation.

Media displacement occurs due to superior content, technical benefit, and cost efficiency (Lin, 2001). Television displaced radio because of television's visual capability. Lin (1994) argued that audiences abandon old technologies and replace them with new ones when the latter is considered more functionally desirable. Thus, new media such as the Internet will displace functionally similar traditional media if people perceive that it is superior in function or content, less costly, or more convenient (Lin, 2001). For example, as a result of television, people discontinued listening to the radio, movie theater attendance declined, and general-interest magazines ceased publication. Barnett, Chon, Park, and Rosen (2001) suggest that the Internet is replacing the telephone because of its functional advantages of data transmission, asynchrony, and cost.

Lazarsfeld (1940) studied broadcasting's impact on print media based on the assumption that new media would lead to reduced use of old media. Increased television viewing has led to reductions in time spent going to the movies, listening to radio, and reading fiction (Coffin, 1955, Bogart, 1956; Parker, 1963). Also, Robinson (1981) reported that individuals' use of television reduced the time spent with other media and social activities. Kayany and Yelsma (2000) also found displacement effects of online media on the use of television, newspapers, and family conversation.

Many studies, however, provided scant support for time displacement due to the functional equivalence of other media. Internet users do not appear to experience displacement (DiMaggio et al., 2001). Studies have shown *media supplementation* in which the adoption of new media increases the time spent with older technologies. This depends on whether the information technology is "functionally similar" to those already in use (Atkin, 1993; Atkin et al.,

1998; Kang & Atkin, 1999; LaRose & Atkin, 1992). Media technologies rarely become extinct; they typically endure by improving or updating their content offerings or delivery system (Lin, 2001).

As a supplemental media technology, the Internet, even though it is more convenient and affordable, provides an additional channel of communication to telephone, face-to-face contact, and the traditional mass media. Dutton et al. (1987) found that computer use was not related to television viewing. Atkin et al. (1998) and Busselle et al. (1999) reported that traditional media use was not significantly related to Internet use. Further, Robinson and Kestnbaum (1999) found that Internet users more frequently use other media or engage in social activities than nonusers. They read more literature, attended more arts events, went to more movies, and watched and played more sports. However, the relationship between use of new media and old media differs by user characteristics. Robinson et al. (2002) reported that Internet use was associated with greater use of print media among users who had been early adopters, while this relationship had disappeared among later Internet users.

In a recent study Nam and Barnett (2010) examined longitudinal trends in Korean use of domestic and international mail, domestic and international telephone calls, telex, mobile telephones, television, and the Internet to determine the various media's displacement and supplement effects. The results show that international mail, domestic telephone, and telex can be best described by a quadratic equation indicating their use is being discontinued, while the trends for domestic mail and international telephone calls show exponential growth or adoption. Correlations indicate that the new media have displaced international mail, domestic telephone and telex, yet they do not substitute for domestic letters or international telephone calls. Finally, television, mobile telephones, and the Internet are at the exponential or logistic growth stage of diffusion and supplement each other. Thus, one can conclude that the various media interact in a variety of ways in the diffusion process. The new innovative media can replace existing media by performing the same functions as the old media, or they can complement the current technologies by supplementing their role in the communication process.

Summary and Conclusions

The traditional models of diffusion described above focused on initial adoption. They assumed that the probability of adoption of a new product or technology at time *t* given that it has not yet been adopted depends on two forces, one of which is independent of the number of previous adopters and the other would be positively influenced by previous adopters (Bass et al., 1994). A bell-shaped curve is created since the new product expands in numbers during the first half of the diffusion process and thereafter declines in numbers during the ensuing time periods. The generational diffusion model develops an understanding of the demand dynamics and interrelationships between generations of innovations (Norton & Bass, 1987). There are three phases of generational diffusion. The first phase is the introduction and development of uses for the first generation, leading to a rapid rise in use. The S-shaped curve represents this single generation and is proportional to the distributed cumulative function of the adoption rate. It does not allow innovation substitution. The second is the introduction of the second-generation innovation, which causes the first to become negative. The new innovation is a replacement. It acts as a substitute for the initial product, practice, or idea. The third phase occurs when the second generation has saturated the substitution potential. At this point, the first generation becomes obsolete as use reaches zero or its eventual use level (Johnson & Bhatia, 1997). The Barnett and Fink models describe the entire trajectory of the first-generation innovation. Information technology products and services are typically released in successive generations that diffuse in wave-like patterns (Bass & Bass, 2004). The generational diffusion model is graphically represented in Figure 8.

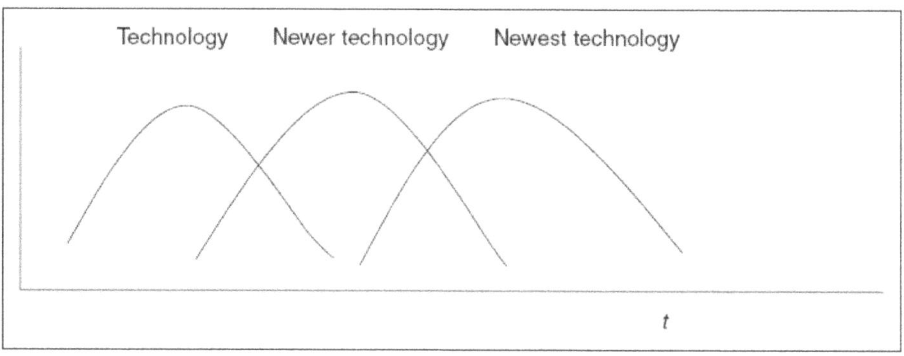

Figure 8: General Diffusion Model.

Barnett, et al., (1993) found a repeating pattern whereby the introduction of simple innovations altered the adoption of television, a complex innovation (Barnett, 1988). They reported a cycle of about 13.8 years in length, which may be characterized as a product development cycle. In the mid-1950s, color television was introduced. Cable television began its diffusion in the early 1970s. The video tape recorder was diffused in the 1980s, which allowed people time shift, i.e., to delay viewing of broadcast programs (Levy & Fink, 1984). The late 1990s brought the Internet and more recently high-definition (HDTV) and flat-panel television have been diffused. All of these simple innovations represent newer generations of the more general innovation that affected how our society views television.

The generational model suggests that future research should examine the diffusion of innovations using Fourier analysis to determine innovation lifecycles (Barnett, Cho & Choi, 1993) and autoregressive models to examine the factors impacting the displacement or supplemental effects of later generations of innovations (Barnett, Chang, Fink & Richards, 1991). One point that should be clear from this chapter is that the traditional S-shaped curve is antiquated for describing how innovations are adopted over time and that alternatives exist that more precisely the diffusion process.

References

Abrahamson, E., & Rosenkopf, L. (1996). Social network effects on the extent of innovation diffusion: A computer simulation. *Organization Science*, 8(3), 289–309.

Allport, G. W., & Postman, L. (1947). *The social psychology of rumor*. New York: Henry Holt.

Atkin, D. (1993). Adoption of cable amidst a multimedia environment. *Telematics and Informatics*, 10, 51–58.

Atkin, D., Jeffres, L., & Neuendorf, K. (1998). Understanding Internet adoption as telecommunications behavior. *Journal of Broadcasting & Electronic Media*, *42*(4), 475–490.

Bandura, A. (1977). *Social learning theory*. Englewood Cliffs, NJ: Prentice Hall.

Barnett, G. A. (1988). An associational model for the diffusion of complex innovations. In G.A. Barnett & J. Woelfel (Eds.), *Readings in the Galileo system: Theory, methods and applications* (pp. 55–74). Dubuque, IA: Kendall/Hunt.

Barnett, G. A., Chang, H. J., Fink, E. L., & Richards, W. D. (1991). Seasonality in television viewing: A mathematical model of cultural processes. *Communication Research*, *18*, 755–772.

Barnett, G. A., Cho, S. & Choi, Y. (1993, June). Seasonality in television viewing: A dampened oscillation model. *Cycles*, 137–141.

Barnett, G. A., Chon, B. S., Park, H. W., & Rosen, D. (2001, April). A network analysis of international Internet flows. Paper presented to the Sunbelt Social Networks Conference. Budapest, Hungary.

Barnett, G. A., & Fink, E. L. (2008). Impact of the Internet and scholar age distribution on academic citation age. *Journal of the American Society for Information Science*, 59(4), 526–534.

Barnett, G. A., Fink, E. L., & Chang, H. J. (1988, May). The adoption and discontinuance of motion picture attendance and monochromatic television: Further test of a mathematical model. Paper presented to the International Communication Association, New Orleans.

Barnett, G. A., Fink, E. L., & Debus, M. B. (1989). A mathematical model of academic citation age. *Communication Research*, *16*, 510–531.

Barnett, G. A. & Siegel, G. (1988). The diffusion of computer assisted legal research systems. *Journal of the American Society for Information Science*, *39*, 224–234.

Bass, F. M. (1969). A new product growth model for consumer durables. *Management Science*, *13*(5), 215–227.

Bass, P.I., & Bass, F.M. (2004). IT waves: Two completed generational diffusion models, *Bass economics*, 1–33. Retrieved from http://www.bassmodelinstitute.org/Research.aspx July 2007.

Bass, F. M., Krishnan, T. V., & Jain, D. C. (1994). Why the Bass model fits without decision variables. *Marketing Science*, *13*(3), 203–223.

Beninson-Germano, M., Barnett, G. A., Dinkelacker, J. W., & Downing, K. A. (1988). Automated ordering in the pharmaceutical industry. In G. A. Barnett & J. Woelfel (Eds.), *Readings in the Galileo System* (pp. 265–294). Dubuque, IA: Kendall/Hunt.

Bogart, L. (1956). *The age of television: A study of viewing habits and the impact of television on American life*. New York: Ungar.

Burt, R.S. (1987). Social contagion and innovation. *American Journal of Sociology*, *92*, 1287–1335.

Busselle, R., Reagan, J., Pinkleton, J. B., & Jackson, K. (1999). Factors affecting Internet use in a saturated-access population. *Telematics and Informatics*, 16, 45–58.

Chang, H. J., & Barnett, G. A. (1991, May). An examination of diffusion of innovations: A mathematical test of five adoption and disadoption processes. Paper presented to the International Communication Association.

Coffin, T. (1955) Television's impact on society. *American Psychologist*, 10, 630–641.

Coleman, J. S. (1964). *Introduction to mathematical sociology*. New York: Free Press.

Coleman, J. S., Katz, E., and Menzel, H. (1966). *Medical innovations: A Diffusion study*. Indianapolis: Bobbs-Merrill.

Danowski, J. A., Gluesing, J., & Riopelle, K. (2009, May). ICT's effects on diffusion curves, e-mail networks, and semantic networks in an organization: The emergence of herding, bandwagons, and information cascades with less interpersonal communication. Paper presented to the International Communication Association, Chicago.

DiMaggio, P., Hargittai, E., Neuman, W.R., & Robing, J.P. (2001). Social implications of the Internet. *Annual Review of Sociology*, *27*, 307–336.

Dodd, S. C. (1955). Diffusion is predictable: Testing probability models for laws of interaction. *American Sociological Review*, *20*, 392–401.

Dutton, W. H., Rogers, E. M., & Jun, S. H. (1987). Diffusion and social impacts of personal computers. *Communication Research*, *14*(2), 219–250.

Hamblin, R. L., Jacobsen, R. B., & Miller, J. L. L. (1973). *A mathematical theory of social change*. New York: John Wiley.

Jackson, M. O. (2008). *Social and economic networks*. Princeton, NJ: Princeton University Press.

Jeffres, L., Atkin, D., & Neuendorf, K. (1995). The impact of new and traditional media on college student leisure preferences. *World Communication*, *24*, 67–73.

Johnson, W. C., & Bhatia, K. (1997). Technological substitution in mobile communications. *The Journal of Business and Industrial Marketing, 12*(6), 383–399.

Kang, M. E., & Atkin, D. (1999) Exploring the role of media uses and gratification in multimedia cable adoption. *Telematics and Informatics, 16*, 59–74.

Kang, N., & Cheng, M. M. (1989, August). The future of movie attendance: From the perspective of diffusion theory. Paper presented to the Association for Education in Journalism, Washington.

Kayany, J. M., & Yelsma, P. (2000). Displacement effects of online media in the socio-technical contexts of households. *Journal of Broadcasting and Electronic Media, 44*(2), 215–230.

LaRose, R., & Atkin, D. (1992). Audiotext and the re-invention of the telephone as mass medium. *Journalism Quarterly*, 69, 413–421.

Lazarsfeld, P. F. (1940). *Radio and the printed page*. New York: Dell, Sloan and Pearce.

Levy, M. R. & Fink, E. L. (1984). Home video recorders and the transience of television. *Journal of Communication, 34*, 56–71.

Lin, C. A. (1994). Audience fragmentation in a competitive video marketplace. *Journal of Advertising Research, 34*(6), 1–17.

Lin, C. A. (2001). Audience attributes, media supplementation, and likely online service adoption. *Mass Communication & Society*, 4(1), 19–38.

Mahajan, V., Muller, E., & Bass, F. M. (1990). New product diffusion models in marketing: A review and directions for research. *Journal of Marketing*, 54, 1–26.

Mahajan, V., & Peterson, R. A. (1985). *Models for innovation diffusion*. Beverly Hills, CA: Sage.

Nam, Y., & Barnett, G. A. (2010). Communication media diffusion and substitutions: Longitudinal trends from 1980 to 2005 in Korea. *New Media & Society*, 12(4), 1–19.

Norton, J. A., & Bass, F. M. (1987). A diffusion theory model of adoption and substitution for successive generations of high technology products. *Management Science* 33(9), 1069–1086.

Parker, E. (1963). The effects of television on public library circulation. *Public Opinion Quarterly, 27*(4), 578–589.

Robinson, J. (1981). Television and leisure time: A new scenario. *Journal of Communication 31*, 120–130.

Robinson, J. P., & Kestnbaum, M. (1999). The personal computer, culture and other uses of free time. *Social Science Computer Review* Summer, 209–216.

Robinson, J. P., Kestnbaum, M., Neustadtl, A., & Alvarez, A. (2002). Information technology and functional time displacement. *IT & Society 1*(2), 21–36.

Rogers, E. M. (2003). *Diffusion of innovations (4th Ed.)*. New York: Free Press.

Rogers, E. M., & Kincaid, D.L. (1981). *Communication networks: Toward a new paradigm for research*. New York: Free Press.

Rosengren, K. E. (1987). The comparative study of news diffusion. *European Journal of Communication, 2*, 227–255.

Valente, T. W. (1994). *Network models of the diffusion of innovations*. Cresskill, NJ: Hampton.

Vishwanath, A., & Chen, H. (2006). Technology clusters: Using multidimensional scaling to evaluate and structure technology clusters. *Journal of the American Society for Information Science, 57* (11), 1451–1460.

· 6 ·

The Revolution in Diffusion Theory Caused by New Media

James A. Danowski, Julia Gluesing & Ken Riopelle

Introduction

From the face-to-face discussions with farmers about seed corn in the 1940s grew the famous S-shaped diffusion curve. Most scholars still unquestioningly teach the classic curve, despite the fact that face-to-face communications are no longer nearly as important among a range of new media. With new media comes a different set of assumptions about the key variables that grounded the old S-shaped curve. A new form of curve emerges: the convex curve, one shaped similarly to the trajectory of a rocket launched into low-earth orbit. This chapter presents the rationale underlying this revolution in diffusion theory wrought by new media technologies. Empirical evidence comes from a study of three innovation concepts at two levels of analysis, individual and organizational units, innovations communicated mainly through e-mail and web pages about automobile engineering innovations in an organization. Herds form when people see in new media that an innovation is going to develop, and they adopt without discussion or little interaction with other individuals about the innovation. The more slowly growing S-shaped curves still explain certain types of innovation diffusion, but S-shaped curves' relegation to a more circumscribed position in the theoretical space about innovation is revolutionary and a pattern likely to continue to expand in scope.

Problem Statement

In his sweeping assessment of research across multiple disciplines, the National Science Foundation's co-director of Human-Centered Computing, William Bainbridge (2004), states that "understanding semantic approaches to cultural systems would enable the engineering of culture, to accelerate the rate of invention in valued fields, and to learn how to fit difficult species of technology into harmoniously functioning ecologies. This is one of the profound revolutions that technological convergence would create" (p. 174).

The proposed research illustrates one example of Bainbridge's profound semantic revolution, the over-time study of e-mail content using semantic-network analysis of full texts. This is done to better understand the diffusion of information and communication technologies (ICTs) in organizations. A focus on message content can be considered the primary distinguishing feature of the discipline of communication compared with other social sciences. The next most central focus in communication studies is focus on the networks of who communicates with whom as messages are exchanged. Moreover, it is considered important to analyze messages and interpersonal networks *over time*. The most developed theory about communication over time is diffusion of innovations theory (Rogers, 1962, 1971, 1983, 1995, 2003). The 2003 edition alone is cited in over 15,000 scholarly works. Nevertheless, comparatively little research has studied diffusion *within* organizations. Most diffusion research that involves organizations looks at them as the unit of analysis as they on the whole adopt innovations among other organizations. In contrast, the current study examines the diffusion of an interconnected set of six innovations within an organization.

In conducting the proposed multi-level investigation, the current study is grounded in a widely useful theoretical and methodological approach to a variety of research problems: network analysis. Showing the wide scope of a network perspective, Barabasi (2002) asserts that everything in human experience is linked to everything else through networks. In communication studies, networks have been analyzed spanning levels of analysis (Castells, 1996), some examples including from nations as nodes linked by their annual telephone call volumes (Barnett, 2001; Danowski, 2000), to networks among cultures constructed from cross-posted discussion list messages (Choi & Danowski, 2002), to networks of information industry mergers and acquisitions (Danowski & Choi, 1998; Chon, Choi, Barnett, Danowski & Sung-Hee, 2003), to networks of sub-disciplines according to co-membership in divisions of a professional association (Barnett & Danowski, 1992), to individuals' e-mail networks (Danowski & Edison-Swift, 1985; Diesner, Frantz & Carley, 2005), to news story word networks (Danowski, 1993a; Corman, Kuhn, McPhee & Dooley, 2006), to concept cooccurrences across discussion forum posts (Danowski, 1982), to individuals' cognitive networks from open-ended survey responses (Rice & Danowski, 1991, 1993; Danwoski, 1993b), studying such things as perceptions of voice mail.

Network Theory and Network Analysis

It has become widely recognized in this decade that the network perspective reflects the fundamental structure of social processes. Borgatti and Foster (2003) show the exponential growth curve for studies on social networks in Sociological Abstracts, reviewing nearly 200 studies of social networks and organizations at both the inter- and intra-organizational levels. Organizational and communication scholars have addressed the emergence of knowledge networks in global organizations and their relationships with information-technology-driven organizations (Contractor & Eisenberg, 1990). Researchers have been developing sophisticated computational simulation models for testing hypotheses about networks and information diffusion, changes in individual and group knowledge and interaction networks, the dynamics of cultural influence networks, and how shared beliefs evolve, focusing on their co-evolution with information technology (Carley & Krackhardt, 1996; Carley, 1996; Contractor, Zink & Chan, 1998; Harrison & Carroll, 2002). *The New York Times* (Eakin, 2003) is even publishing articles about the popularity of network theory, and there are best-selling books on the topic (Gladwell, 2000; Barabasi, 2002; Buchanan, 2002; Johnson, 2001; Watts, 1999, 2003; Strogatz, 2003). Physicists have conducted numerous studies of networks and various social practices, modeling them in high-order mathematical network terms (Newman, 2002).

Network theory, as it has been applied to the study of human behavior and relationships, is comprised of multiple theoretical approaches. Monge and Contractor (2001, 2003) state that there are 10 families of theories that have been used to explain the emergence, maintenance, and dissolution of communication networks in organizations. With a long tradition in sociology, organizational theory, and anthropology, network analysis is a form of structural analysis with theory and methods intimately linked (Monge & Contractor, 2001; Rogers & Kincaid, 1981; Bernard & Ryan, 1998; Borgatti & Foster, 2003). The analysis technique is most often used to discover the pattern of interpersonal communication in a social system by determining who talks to whom, and investigating structural and relationship properties of networks (Valente, 1995, 1996; Cross, Borgatti & Parker, 2002). Monitoring emerging networks identifies where greater leverage can be gained for channeling diffusion resources (Cotrill, 1998; Carley, 1995). Although inter-organizational networks are not of direct focus in the current research, there is a stream of studies that investigate inter-organizational network predictors of organizational adoption of innovations (Davis, 1991; Haunschild, 1993; Palmer, Jennings & Zhou, 1993; Powell, Koput & Smith-Doerr, 1996; Geletkanycz, Boyd & Finkelstein, 2001; Gulati & Westphal, 1999; Westphal, Seidel & Stewart, 2001).

Positional, relational, and cultural traditions in network analysis originate from the classic, seminal structural concepts developed in sociology and anthropology by theorists such as Durkheim and Radcliffe-Brown, and from organizational theorists such as Weber, Parsons, and Homans (Monge & Contractor, 2001). In organizations, this approach sometimes focuses on formal structures. Other researchers have used network analysis to describe the communication links in organizations regardless of formal position. This network structure is often referred to as the informal network and places emphasis on the dynamic and constantly changing nature of the links, based on repetitive patterns of message flow or the network's emergent properties. From the cultural perspective, networks are seen as socially constructed through a series of communicative interactions in which shared meanings emerge that constrain and enable interactions. Work done from this tradition has focused on semantic networks or word networks to examine message content and the structure of meanings in communicative interactions (Danowski, 1982, 1993b; Monge & Eisenberg, 1987).

Networks are increasingly seen as patterns of communication (Borgatti, 2008) about a variety of topics such as friendship or social interaction, technical or professional advice, and the sharing of new ideas or techniques (Scott, 1991, 2000; Wellman, 1983). Typically, networks are constructed by asking respondents to name others with whom they communicate, sometimes specifying a particular topic. These traditional survey research methods for gathering network data have restricted the ability of researchers to examine networks dynamically, or to investigate large networks that cross social systems.

In contrast, the current study uses methods that tap the capabilities of information technology infrastructure to overcome these restrictions in the main data collection phases by analyzing organizational e-mail content and e-mail distribution networks gathered by ICT procedures that recover historical information as well as collect real-time data. Our study focuses on the organizational communication networks enabled by information technology infrastructure. We research the diffusion of innovation in IT-supported networks of practice. Of particular relevance to this research are recent studies of the social construction of innovation networks. Poole and DeSanctis (1990) have examined how actors and structures in a social system influence each other in a recursive relationship. In a longitudinal study conducted at a U.S. public works department, the duality of this relationship was empirically validated using the output from simulation techniques in comparison with actual network evolution (Contractor, Whitbred, Fonti, Hyatt, O'Keefe & Jones, 2000). Harrisson and Laberge (2002) explored the process of diffusion of a socio-technical innovation among workers of a large microelectronics firm. Network analysis revealed how innovation is constituted and the communicative form it takes by tracing the chain of arguments and responses. Burkhardt and Brass (1990) demonstrated in their study how the diffusion of an innovation altered the network structure based on knowledge and information individuals possessed about the innovation. Torenvlied and Velner (1998) investigating resistance to the introduction of ISO quality standards in a transport company found that contagion of resistance in an informal trust network is a significant barrier in diffusion of innovations.

Network Models of Diffusion

Networks are important to the diffusion of innovation (Debresson & Amesse, 1991) because they posit that the ties between individuals influence the spread of an innovation. Most diffusion models are contagion/epidemic/cohesion/relational models where information about innovation is passed from one person to another through direct contact. Valente (1995, 2005) identified only six studies that exist in the public domain that utilized network models of the diffusion of innovation with both network data and time of adoption data. He reanalyzed data from three of the studies to demonstrate how relational network models, structural network models, threshold and critical mass models aid our understanding about how ideas, products, and opinions "take off" and spread with varying speed through a social system. Valente (1995) conceptualized a network threshold model that is both relational and structural and provides a more accurate measure of a person's innovativeness. He calls out the need for more network and diffusion research that measures adoption over time while collecting network data (Valente, 2005).

Network Growth and Decay

Most research treating diffusion of innovations from a network perspective has considered networks as static. Nevertheless, as Borgatti and Foster (2003) point out, the research on change in network structure is rich in examining the evolution of group structure, including empirical investigations of network change (Barnett & Rice, 1985; Burkhardt & Brass, 1990; Burt, 2000; Shah, 2000; Barnett, 2001), and general mathematical models of change (Doreian & Stokman, 1997; Snijders, 2001), as well as agent-based simulation studies reviewed by Macy and Willer (2002). Nevertheless, because of the difficulty of simultaneously collecting network data and time of adoption data, the testing of network threshold and related models of diffusion have remained limited.

In short, while network approaches to diffusion have been infrequent, even less researched has been diffusion in online organizational environments. The broad question of the current study is: Do different diffusion models emerge within the online organization compared with the traditional societal diffusion context?

Diffusion of Innovations Theory

Research on the diffusion of innovations spans almost six decades (Griliches, 1957; Nyblom, Borgatti, Roslakka & Salo, 2003) and includes more than 5,000 studies. No other field in the behavioral and social sciences represents more effort by more scholars in more nations (Rogers, 2003). An innovation can refer to new knowledge, to new technologies such as information technologies, product improvements or manufacturing technologies, or to a new process for doing work in organizations. While there is a large body of extant research about innovation based on product or process life cycle (Utterbeck, 1994; Fine, 2001), this study is grounded in the theoretical and methodological traditions in communication and social network research.

There are a few diffusion studies that look at organizational adoption of ICTs (Markus, 1987, 1990; Mahler & Rogers, 1999). For example, researchers studying the adoption of enterprise resource planning software in Europe found that factors affecting late adoption differed from those explaining early adoption (Waarts, van Everdingen & Hillegersberg, 2002).

Our focus on measuring diffusion via ICT data does not suggest that face-to-face interaction is unimportant to diffusion. Following the findings of

Haythornthwaite and Wellman (1998), who reported that social network data on media use among members of a co-located research group showed that pairs with closer ties used more media to communicate, we assume that to some extent ICT-based networks are correlated with face-to-face network structures, although in the current study we hypothesize that interpersonal communication is less important to diffusion in heavily used ICT organizational environments than it is in less mediated social systems.

One of the most robust findings from the diffusion of innovation literature is the S-shaped cumulative adoption curve (Rogers, 1995), called in the marketing literature the "Bass Model" (Bass, 1969, 1980). The curve results from a simple plot of the cumulative number of adopters of an innovation on the Y-axis against some fixed period of time along the X-axis. Henrich (1999) names a second kind of curve, r-shaped or r-curves. These begin convexly with a maximum growth rate and then slowly taper off toward equilibrium; the difference between the two kinds of curves is shown in Figure 1.

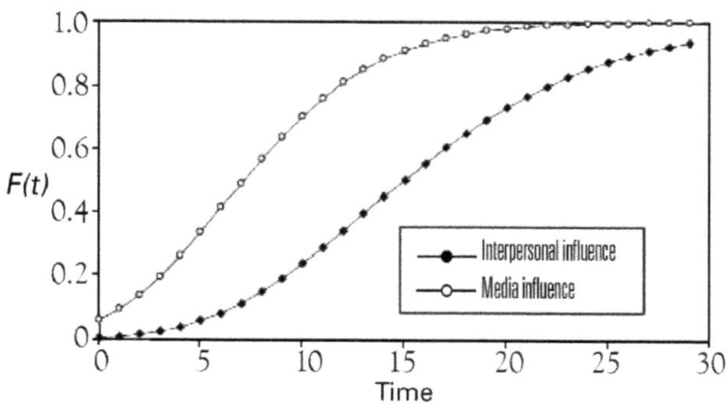

Figure 1

Scholars such as Hägerstrand (1952), Bowden (1964, 1965), Bass (1969), Hanneman (1969), Carley and Svoboda (1996), and Henrich (1999) have developed computational models to simulate the diffusion of innovations under various conditions and parameters that produce these curves as a graphic summary of the diffusion process over time. Most such curves are S-shaped.

In contrast, Rosenkopf and Abrahamson (1999) report "bandwagon" production factors resulting in more r-shaped or convex diffusion curves. These predictors include: (1) greater number of media messages, (2) more ambiguity of innovation efficiency/effectiveness, (3) more messages about number of adopters, and (4) more messages about status of adopters. We expect higher values for these variables in organizational environments when diffusion is primarily via ICT, in the current study via e-mail. E-mail "media" would seem to typically produce large numbers of messages about innovations in organizations. New product innovations, such as those studied in the current research—new car features in the concept stage—have associated uncertainties. There is less common understanding of the overall concept of the innovations, how these will be engineered, how feasible it will be to integrate them across different car platforms, and how much they will cost to build. These uncertainties link with Rosenkopf and Abrahamson's second predictor of convex diffusion curves. In the current study these aspects of the innovations are quite ambiguous in terms of efficiency and effectiveness. In fact, one innovation was eventually terminated because of negative performance in these areas.

Consistent with the third predictor, e-mail provides information about the number of adopters through the CC: function. It is the norm in the organization to CC: members of the work group, key people in other work groups, and members of the steering committee for the innovations. In this way individuals have increasing messages about the number of adopters as more e-mail about the innovations is produced when individuals begin to adopt the innovations and e-mail about them. The fourth predictor, adopter status, is highly visible because widely used e-mail "signatures" provide clear status information. Thus, all four of the conditions Rosenkopf and Abrahamson (1999) specify for convex diffusion curves are met in the organizational innovations studied in the current investigation.

Others theorize about the causes of rapid-growth convex curves for diffusion in terms of herd behavior (Geroski, 2000) leading to information cascades. Herd behavior (Choi, 1997) is taken from animal behavior observations. Once lead members of the herd begin to stampede in a particular direction, the other members quickly follow suit with no knowledge of environmental conditions except that the lead herd animals are moving. In a stock market, for example, once enough investors sell or buy stocks rapidly, other investors follow in a herd-like manner with no information other than sufficiently large numbers indicating other investors are making a move, making it unquestionably important to mimic. This makes the resulting adoption curve such that it has a very rapid growth over time followed by a tapering off as laggards follow.

We propose that the underlying interpersonal communication network is not dense and interlocking throughout this process, but instead is highly constrained, which, in network terms, manifests as simple center/periphery structures or highly radial networks in which the contacts of individuals do not communicate with one another. In information theory terms the former type of network structure is entropic while the latter type is negentropic. Negative entropy is the opposite of entropy or equal distributions of events. In contrast, negative entropy is a result of constraints on all possible events. The more constrained a situation, the higher the negative entropy. In information theory, negative entropy means the same as information, the reduction of uncertainty or of entropy. In a network, if all nodes were completely connected to all other nodes, this would be the most entropic pattern. If, on the other hand, the network had a central hub with contacts emanating from it, which were not connected to each other, this would be a pattern showing high negative entropy. For short, we refer to negative entropy as negentropy.

It is reasoned that the primary cause of the logistic curve is face-to-face interpersonal communication in which individuals embody a contagion or epidemic process where they discuss with others their experience with the innovation and decide whether or not to adopt it. In the famous hybrid seed corn diffusion example (Griliches, 1957), it took years for the interpersonal contagion process to unfold into the S-shaped curve. This appears to be a generalizable notion (Dixon, 1980), consistent with Geroski's (2000) theory that direct, unmediated interpersonal contact as the basis for passing of innovation information results in an S-shaped curve.

Geroski theorizes that the contrasting r-shaped modified exponential curves occur because of information cascades. When individuals adopt an ICT innovation without having to go through the learning curve themselves, but make their adoption decision based on the experiences of the early adopters, time of adoption speeds up compared with the contagion-based S-curves. Geroski posits three phases to such information cascade or bandwagon adoption: (1) initial choice of early adopters between technology variant A and B, (2) lock-in (Arthur, 1989) on A, (3) bandwagon or information cascade effect. Bass (1969) proposed early on (see Figure 1) that some innovations rely heavily on interpersonal communication and therefore have the S-shaped curve while other innovations involve more use of the media for innovation-relevant messages, resulting in the convex r-curves. Rosenkopf and Abramson's (1999) information cascades theorizing provides more specific reasons for its occurrence and does so in communication terms.

With the ubiquity and sophistication in logging of e-mail ICTs, researchers can take more advantage of automated monitoring of communication to investigate organizational innovations. Although this has been discussed for some time in the literature (Danowski, 1982, 1988) such methods have diffused slowly.

Additionally, this research addresses five shortcomings of traditional diffusion research. Those studies: (1) are piecemeal, descriptive, and retrospective, (2) relate a single variable to diffusion, (3) collect data from error-rich self-reports, (4) rarely include both time of adoption and network data; Valente (1995, 2003) identified only six such studies (e.g. Danowski, 1976), and (5) give little theoretical attention to why ICTs may change diffusion processes.

The logistic function defines the S-shaped curve. It can be defined as:

$$P(t) = \frac{1}{1+e^{-t}} = \frac{1}{1+\exp(-t)} = (1+\exp(-t))^{-1}$$

where the variable *P* denotes a population or sample and *t* is *time*.

Hypotheses

Based on the literature and reasoning discussed, we propose the following hypothesis:

> H: Diffusion curves for IT-based innovations will be convex rather than S-shaped.

Methods

Organization and Innovations Studied

This study was conducted among an American automobile company's product engineering staff across its global network. The innovations were initially a package of two sets of products with a single overall name that was created to be a new feature across a range of the company's vehicles. The number of engineers involved was approximately 2,000.

ICT Studied

Communication about the innovations occurred in a variety of modes, but the most tractable was e-mail about the innovation. Monitoring all relevant electronic mail over time overcomes limitations of cross-sectional, self-report data, which includes considerable errors introduced by respondent memory processes and their ability to report only gross summary features of their communication messages about an innovation. To avoid such problems, this study created a procedure by which the over time e-mail content of individuals was monitored both historically and in real time over a two-year period. The organization exclusively used Microsoft Outlook for e-mail with approximately 400 servers with up to 4,000 users each. The first step in establishing the monitoring procedure was to create a new Outlook rule that a participant would apply once on his Windows PC so that it would search all historical e-mails stored by individuals for key words associated with the innovations and forward these e-mails to a dedicated server—and also forward all relevant e-mails in real time during the course of the study, from January 2005 to December 2007.

A major discussion about deploying this procedure was among the executives and corporate attorneys to ensure confidentiality, respect for the personal privacy of the participants and compliance with the legal systems of employees from different countries. After an in-depth review by the company's lawyers from multiple countries, the process was approved as an opt-in choice, an approach more characteristic of Europe than the USA. In all cases, participation was voluntary and employees could stop participating at any time. Participants were asked to run rules in Microsoft Outlook to enable automatic forwarding of e-mail to a "dummy e-mail address" on a secure server designated to store the study data. Therefore, study participants were fully aware that their e-mail was being monitored for the study. When they composed an e-mail, they also could see the dummy e-mail address in the CC line and delete it if for some reason they did not want a particular e-mail to be part of our study data.

Once the e-mail data collection process was approved, the academic researchers received IRB approval from the two universities at which the PI and Co-PIs for this NSF grant were located based on a procedure to protect privacy used by Danowski and Edison-Swift (1985) to convert all names in the e-mails to numbers known only to the corporate staff who collaborated in this anonymizing activity. While the corporate research team knew the identities of all named individuals in the e-mails, the academic researchers had no knowledge of these identities. IRB applications were submitted for exempt status for the research, meaning that, because no identifying information was known to the academic researchers, the research was exempt from the need to obtain informed consent.

The first analysis step was to deploy the Outlook rule to the project manager's e-mail going back nine years. The academic researchers then conducted a network analysis of the who-to-whom network from these e-mails to identify the most central individuals using Negopy (Richards & Rice, 1981). The program found one large group, evidence of a negentropic center/periphery structure (see Figure 2). Because Negopy computes the geodesic distance scores among all pairs of nodes in each group, in this case one, comparable centrality information was available for all nodes in the network. This information was presented to the corporate researchers who used this information along with other considerations to choose 298 target individuals for the e-mail harvesting.

The 298 targets were initially sent an e-mail requesting their participation and application of the Outlook rules. Thirty-eight chose to voluntarily participate, for a participation rate of 13%. Follow-up by project staff noted that potential users did not wish to have their e-mail monitored in this fashion because of uncertainty around what it would find and how it would be used, even though most were aware that all corporate e-mail was monitored in real time for "illegal" activity words. Nevertheless, we captured non-participant e-mail that included one of the 38 people who participated on their To: From or CC: list on each e-mail. We found that capturing all innovation e-mails, many having long forwarding chains, from only 1% of the 2,000 project engineers was sufficient to capture e-mails of approximately 1,900 individuals exchanging approximately 45,000 e-mails. As a result we constructed a two-year time series of e-mails about the innovations.

Results

Figures 2–7 show the results for the six innovation diffusion curves. SPSS 18 Curve Fitting procedure was used. It does not have the convex shape as one of its fitting options, but we are more interested in whether the logistic curve explains the most variance. The contrasting curve we will use is the cubic curve. There are many different shapes to the cubic curve, and one of them is the convex curve. Once examining the statistical evidence one can look at the shapes of the curves and visually see whether the cubic functions are the particular kind of interest, convex curves. All of the various curves fitted are statistically significant at ($p < .000$), although the variance explained varies. For the adoption of innovation #1 by organizational units, shown in Figure 2, the logistic curve explained 59% of the variance over time, while the cubic form $f(x)=ax^3+bx^2+cx+d$ explained 97%. For the adoption of innovation #1 by individuals, shown in Figure 3, the logistic curve explained 79% of the variance, while the cubic explained 99%. For the adoption of innovation #2 by organizational units, shown in Figure 4, the logistic curve explained 34% of the variance, while the cubic explained 61%. For the adoption of innovation #2 by individuals, shown in Figure 5, the logistic function explained 93% of the variance, while the linear, cubic, and quadratic each explained 97%. For the adoption of innovation #3 by organizational units, shown in Figure 6, the logistic curve explained 72% of the variance while the cubic curve explained 98%. For the adoption of innovation #3 by individuals, shown in Figure 7, the logistic function explained 81% of the variance and the cubic function explained 97%. Combining the statistical results with the visual curve identification, there is consistent evidence of convex r-curves explaining more variance than logistic S-curves, supportive of the hypothesis. At the organizational level, the cubic functions explain nearly six times more variance than at the individual level.

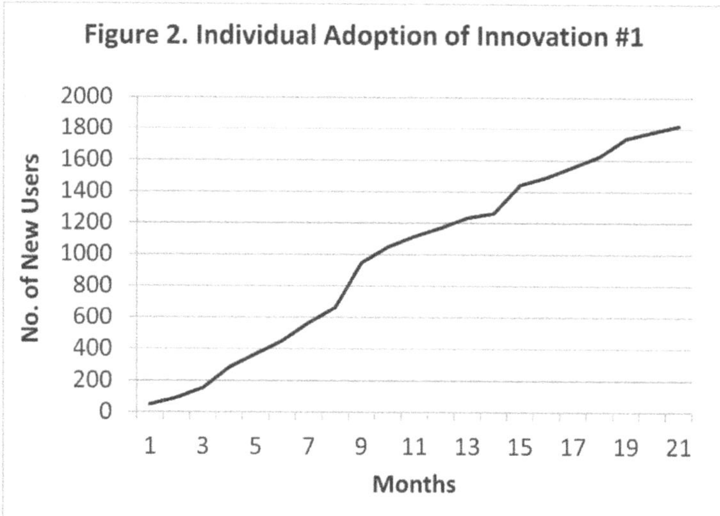

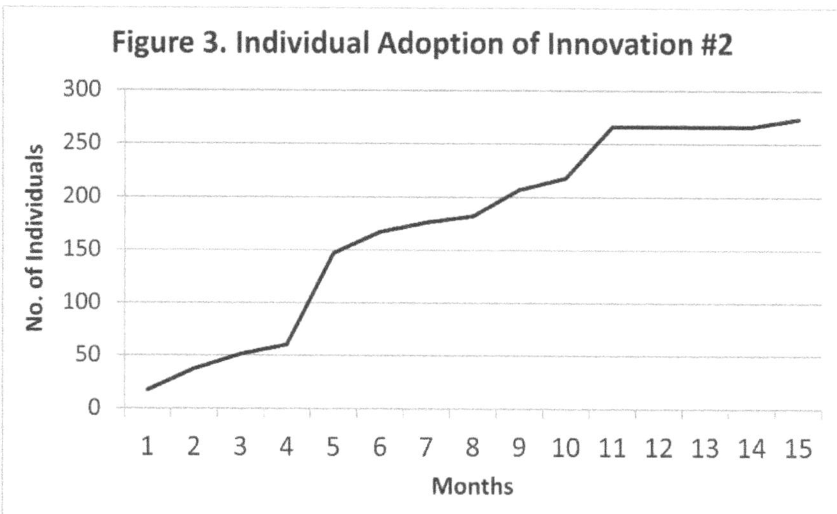

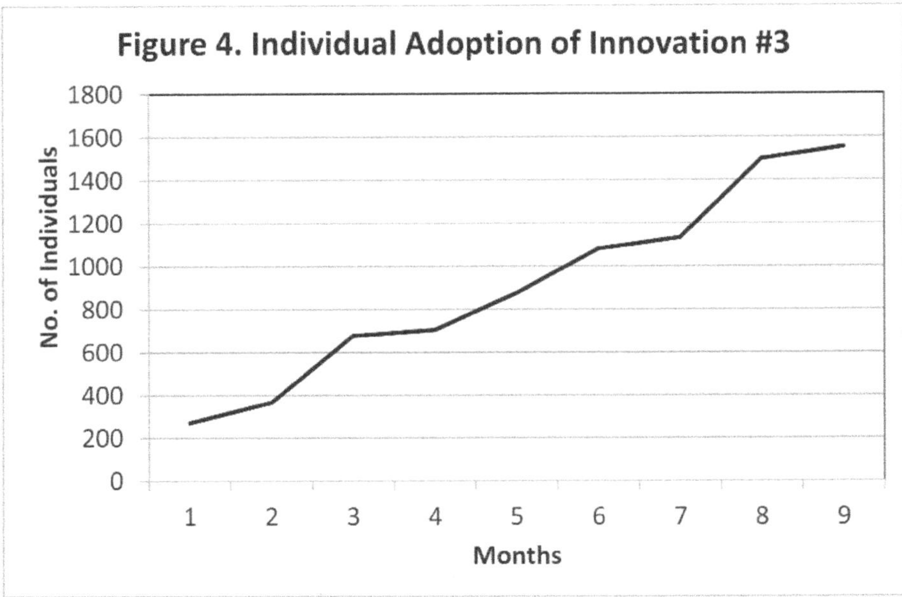
Figure 4. Individual Adoption of Innovation #3
No. of Individuals
1800
1600
1400
1200
1000
800
600
400
200
0
1
2
3
4
5
6
7
8
9
Months

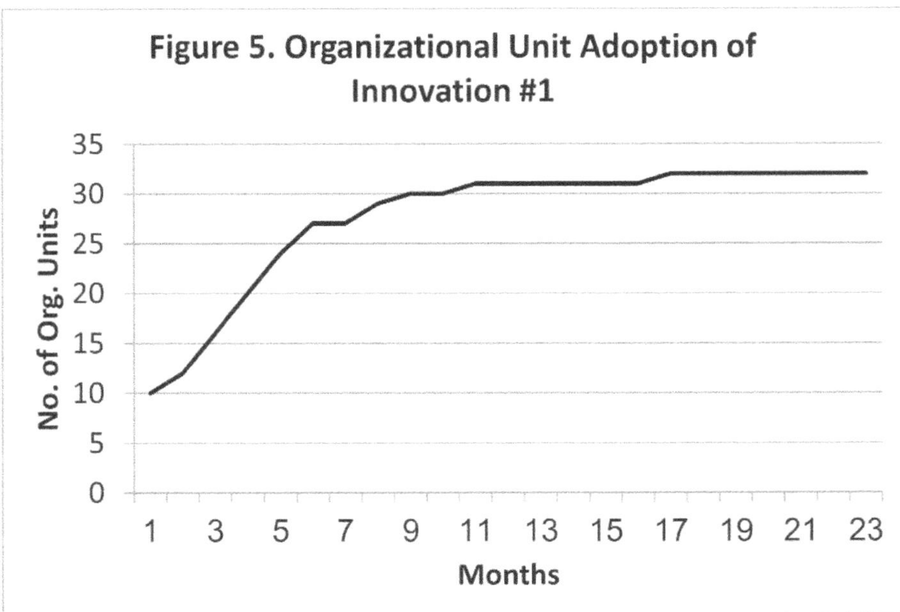
Figure 5. Organizational Unit Adoption of Innovation #1
No. of Org. Units
35
30
25
20
15
10
5
0
1
3
5
7
9
11
13
15
17
19
21
23
Months

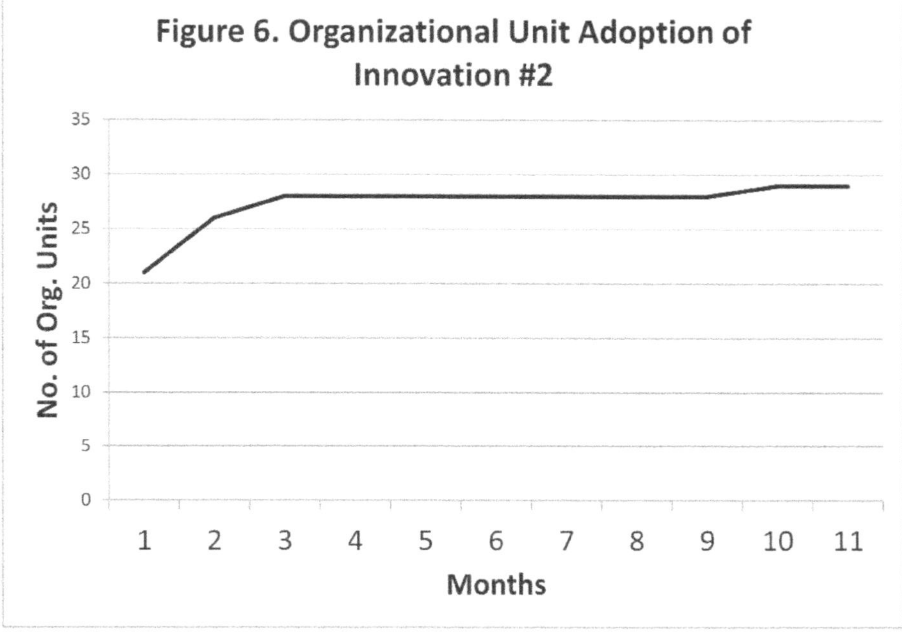
Figure 6. Organizational Unit Adoption of Innovation #2
35
30
25
20
15
10
5
0
No. of Org. Units
1 2 3 4 5 6 7 8 9 10 11
Months

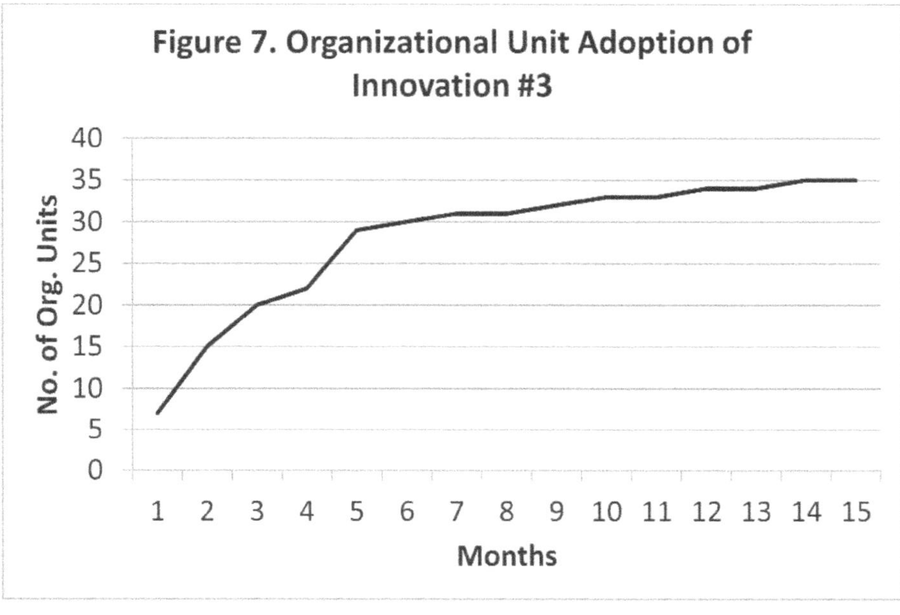
Figure 7. Organizational Unit Adoption of Innovation #3
40
35
30
25
20
15
10
5
0
No. of Org. Units
1 2 3 4 5 6 7 8 9 10 11 12 13 14 15
Months

Discussion

This study found support for the hypothesis that innovations diffused through heavy use of ICTs would have a convex r-curve rather than the traditional S-curve. The classic contagion/cohesion models of diffusion of innovations that involve growing chains of interpersonal interactions for an S-curve of adoption to slowly build do not appear. Instead, a rapidly accelerating convex r-curve develops. This represents a major shift in expecting how innovations may diffuse in increasingly ICT-based environments. There is a more rapid ramp-up of adoption without much direct interpersonal communication, rather than the traditional slow-building S-shaped curves due to the need for potential adopters to directly communicate with earlier adopters and one another in deciding whether to adopt. In the ICT-based global space, one can therefore expect more rapid diffusion of more innovations more quickly over time than in the previous pre-ICT world.

The theoretical rationale is that the ICT studied e-mail allows for herd behavior, otherwise known as information cascade effects. This is perhaps because by examining e-mail headers users can see who the early adopters are and their choices about ambiguous innovations in terms of efficiency and effectivness. They also see the onset of the next phase, the lock-in of others on this choice. Once observing this, there is no need to have interpersonal communication in deciding to adopt, and people quickly decide to join herd movement. Reasons include that with e-mail individuals not only see who is adopting what when and their organizational status but see emerging herd movements that they join. In addition, some adoption may be mandated by management. The nearly six times greater convexity at the organizational level may be in part a result of this mandate factor, but another likely factor may be herd observation of early adopters within an organizational unit. In this study, examination of six diffusion curves showed all of them to be convex r-curves and not S-shaped, therefore supportive of the hypothesis. Future research can see to what extent ICT innovations in other organizations follow this pattern and, more widely, whether ICT innovations among large global populations have more convex than S-shaped diffusion curves.

Note

1. This research was supported by the National Science Foundation's Human and Social Dynamics (HSD) Award #SES-527487: Accelerating the diffusion of innovations: A "digital diffusion dashboard" methodology for global networked organizations.

References

Arthur, B. (1989). Competing technologies, increasing returns and lock-in by historical small events. *Economic Journal*, 99, 116–131.

Bainbridge, W. S. (2004). The evolution of semantic systems. *Annals of the New York Academy of Sciences, 1013*, 150–177.

Balcer, Y., & Lippman, S. (1984). Technological expectations and adoption of improved technology. *Journal of Economic Theory*, *34*, 292–318.

Barabasi, A. L. (2002). *Linked: The new science of networks*. New York: Perseus.

Barnett, G. A. (2001). A longitudinal analysis of the international telecommunications network: 1978–1996. *American Behavioral Scientist*, *44*(10), 1638–1655

Barnett, G. & Danowski, J. (1992). The structure of communication: A network analysis of the International Communication Association. *Human Communication Research*, 19, 164–285.

Barnett, G. A., & Rice, R. E. (1985). Longitudinal non-euclidean networks: Applying Galileo. *Social Networks*, *7*(4), 287–322.

Bass, F. M. (1969). A new product growth model for consumer durables. *Management Science*, *15*(5), 215–227.

Bass, F. (1980). The relationship between diffusion rates, experience curves and demand elasticities for consumer durables technological innovations. *Journal of Business*, *53*, 551–567.

Bernard, H. R. & Ryan, G. W. (1998). Text analysis: Qualitative and quantitative methods. In H. R. Bernard. (Ed), *Handbook of methods in cultural anthropology* (pp. 595–646). Walnut Creek, CA: Altamira

Bikhchandani, S., Hirschleifer, D., & Welch, I. (1992). A theory of fads, fashion, custom and cultural change as informational cascades. *Journal of Political Economy*, *100*, 992–1026.

Bikhchandani, S., Hirschleifer, D., & Welch, I. (1998). Learning from the behaviour of others: Conformity, fads and informational cascades. *Journal of Economic Perspectives*, *12*, 151–170.

Borgatti, S. (2008, January). Keynote address to the annual meetings of the International Network for Social Network Analysts, St. Pete Beach, FL.

Borgatti, S. P., Everett, M., & Freeman, L. (2002). *UCINET 6 for Windows: Software for social network analysis*. Harvard, MA: Analytic Technologies.

Borgatti, S. P. & Foster, P. C. (2003). The network paradigm in organizational research: A review and typology. *Journal of Management*, *29*(6), 991–1013.

Borgatti, S. P., & Molina, J. L. (2003). Ethical and strategic issues in organizational social network analysis. *The Journal of Applied Behavioral Science*, *20*(10): 1–13.

Bowden, L. W. (1964). Simulation and diffusion of irrigation wells in the Colorado northern high plains. Paper presented at the Working Conference on Spatial Simulation Systems. Pittsburg.

Bowden, L. W. (1965). Diffusion of the decision to irrigate: Simulation of the spread of a new resource management practice in the Colorado northern high plains. Chicago: University of Chicago, Department of Geography, Res. Paper 97.

Buchanan, M. (2002). *Nexus: Small worlds and the groundbreaking science of networks*. New York: W. W. Norton.

Burkhardt, M. E. & Brass, D. J. (1990). Changing patterns or patterns of change: The effects of a change in technology on social network structures and power. *Administrative Science Quarterly*, 35, 104–127.

Burt, R. S. (1992). *Structural holes: The social structure of competition*. Cambridge, MA: Harvard University Press.

Burt, R. S. (2000). Decay functions. *Social Networks, 22*(1), 1–28.

Cabral, L. (1990). On the adoption of innovations with network externalities. *Mathematical Social Sciences, 19*, 299–308.

Carley, K. M. (1995). Communication technologies and their effect on cultural homogeneity, consensus, and the diffusion of new ideas. *Sociological Perspectives*, 38(4), 547–571.

Carley, K. M. (1996). Communicating new ideas. The potential impact of information and telecommunication technology. *Technology in Society*, *18*, 219–230.

Carley, K. M. & Krackhardt, D. (1996). Cognitive inconsistencies and non-symmetric friendship. *Social Networks*, *18*, 1–27.

Carley, K. M. & Svoboda, D. M. (1996). Modeling organizational adaptation as a simulated annealing process. *Sociological Methods and Research*, *25*, 138–168.

Castells, M. (1996). *The Information Age: Economy, society and culture, Volume I, The rise of the network society*. Oxford: Blackwell.

Choi, J. (1997). Herd behaviour, the penguin effect and the suppression of informational diffusion. *Rand Journal of Economics*, 28, 407–425.

Choi, J. H. & Danowski, J. (2002). Making a global community on the net—global village or global metropolis?: A network analysis of Usenet newsgroups. *Journal of Computer Mediated Communication*, *7*(3). Retrieved from http://www.ascusc.org/jcmc/vol7/issue3/choi.html.

Chon, B. S., Choi, J. H., Barnett, G. A., Danowski, J. A. & Jo, S. H. (2003). A structural analysis of media convergence: Cross-industry mergers and acquisitions in the information industries. *Journal of Media Economics* 16(3), 141–157.

Contractor, N. S. & Eisenberg, E. M. (1990). Communication networks and new media in organizations. In J. Fulk & C. Steinfield (Eds), *Organizations and communication technology*, (pp. 143–172). Newbury Park, CA: Sage.

Contractor, N., Whitbred, R., Fonti, F., Hyatt, A., O'Keefe, B. & Jones, P. (2000). Structuration theory and the evolution of networks. Paper presented at the 2000 Winter Organizational Science Conference, Keystone, CO.

Contractor, N., Zink, D. & Chan, M. (1998). IKNOW: A tool to assist and study the creation,maintenance, and dissolution of knowledge networks. In Ishida, T. (Ed.), *Community computing and support systems: Lecture notes in computer science*, 1519 (pp. 201–217). Berlin: Springer-Verlag.

Corman, S. R., Kuhn, T., McPhee, R. D. & Dooley, K. (2006). Studying complex discursive systems: Centering resonance analysis of communication. *Human Communication Research*, *28*(2), 157–206.

Cotrill, K. (1998). Networking for innovation. *Chemical Week*, *160*(7), 39–41.

Cross, R., Borgatti, S. P. & Parker, A. (2002). Making invisible work visible: Using social network analysis to support strategic collaboration. *California Management Review*, *44*(2), 25–46.

Cross, R., & Parker, A. (2004). *The hidden power of social networks*. Boston, MA: Harvard Business School Press.

Danowski, J. A. (1976). Communication network analysis and social change: Group structure and family planning in two Korean villages. In G. Chu (ed.), *Communication and group transformations for development*. *Communication Monographs No. 2*. (pp. 277–306). Honolulu: East-West Center.

Danowski, J. (1982). A network-based content analysis methodology for computer-mediated communication: An illustration with a computer bulletin board. In Bostrom, R. N. (Ed), *Communication yearbook*, 6, 904–925. New Brunswick, NJ: Transaction.

Danowski, J. (1988). Organizational infographics and automated auditing: Using computers to unobtrusively gather and analyze communication. In G. Goldhaber & G. A. Barnett (Eds.) *Handbook of organizational communication* (pp. 335–384). Norwood, NJ: Ablex.

Danowski, J. (1993a). Automatic narrative generation via statistical content analysis of large-scale textual collections. Paper presented at Conference on Computing for the Social Sciences, National Center for Supercomputing Applications, University of Illinois at Urbana-Campaign, May 18–21.

Danowski, J. (1993b). Network analysis of message content. In W. D. Richards & G. A. Barnett (Eds), *Progress in communication science*, *XII* (pp. 197–221). Norwood, NJ: Ablex.

Danowski, J. (1993c). WORDij: A word-pair approach to information retrieval. *Proceedings of the DARPA/NIST TREC conference* (pp. 131–136). Washington, DC: National Institute of Standards and Technology.

Danowski, J. A. (2000). Arab countries' global telephone traffic networks and civil society discourse. H. Amin, & L. Gher (Eds.) *Civic discourse and digital age communications in the Middle East* (pp. 93-108). Westport, Conn.: Ablex Publishing.

Danowski, J. A. (2009). WORDij [computer program]. Chicago, IL: University of Illinois at Chicago. (Available http://WORDij.net)

Danowski, J. A., & Edison-Swift, P. (1985). Crisis effects on intraorganizational computer-based communication. *Communication Research*, *12*, 251–270.

Danowski, J., & Choi, J.H. (1998). Convergence in the information industries: Telecommunications, broadcasting, and data processing—1981–1996. H. Sahwanny (Ed.) *Progress in Communication Sciences* (pp. 125-150). Norwood, NJ: Ablex.

Davis, G. F. (1991). Agents without principles? The spread of the poison pill through the intercorporate network. *Administrative Science Quarterly*, 36, 583–613.

Debresson, C., Amesse, F., 1991. Networks of innovations. *Research Policy*, *20*, 363–380.

Diesner, J., Frantz, T. L., & Carley, K. M. (2005). Communication networks from the Enron email corpus: "It's always about the people. Enron is no different." *Computational & Mathematical Organization Theory*, *11*, 201–228.

Dixon, R. (1980). Hybrid corn revisited. *Econometrica*, *48*, 145–146.

Doreian, P., & Stokman, F. (1997). The dynamics and evolution of social networks. In P. Doreian & F. Stokman (Eds), *Evolution of social networks* (pp. 1–17). Amsterdam: G & B Pubs.

Eakin, E. (2003). Connect, they say, only connect. *New York Times*, January 25.

Fine, Charles H. (2001). Innovation and economic performance in the automobile industry over the long twentieth century. In R. B. Nelson, B. Steil & D. Victor (Eds). *Innovation and economic performance* (pp. 1–26). Princeton, NJ: Princeton University Press.

Fredrickson, B. L. & Losada, M. F. (2005). Positive affect and the complex dynamics of human flourishing. *The American Psychologist*, 60(7), 678–686.

Geletkanycz, M. A., Boyd, B. K., & Finkelstein, S. (2001). The strategic value of CEO external directorate networks: Implications for CEO compensation. *Strategic Management Journal*, *22*(9), 889–898.

Geroski, P.A. (2000). Models of technology diffusion. *Research Policy*, 29, 603–625.

Gladwell, M. (2000). *The tipping point*. New York: Little Brown.

Gloor, P. (2006). *Swarm creativity: Competitive advantage through collaborative innovation networks*. New York: Oxford University Press

Griliches, Z. (1957). Hybrid corn: An exploration in the economics of technical change. *Econometrica*, 48, 501–522.

Gulati, R. & Westphal, J. D. (1999). Cooperative or controlling? The effects of CEO-board relations and the content of interlocks on the formation of joint ventures. *Administrative Science Quarterly*, 44(3), 473–506.

Hägerstrand, T. (1952). *The propagation of innovation waves*. Lund, Sweden: Lund Studies in Geography, 4.

Hanneman, G. J. (1969). *A computer simulation of information diffusion in a peasant community*. M.A. thesis. East Lansing, MI: Michigan State University.

Harrisson, D. & LaBerge, M. (2002). Innovation, identities and resistance: The social construction of an innovation network. *Journal of Management Studies*, 39(4), 497–521.

Harrison, J. R. & Carroll, G. R. (2002). The dynamics of cultural influence networks. *Computation & Mathematical Organization Theory*, 8, 5–30.

Haunschild, P. R. (1993). Interorganizational imitation: The impact of interlocks on corporate acquisition activity. *Administrative Science Quarterly*, 38(4), 564–592.

Haythornthwaite, C. & Wellman, B. (1998). Work, friendship and media use for information exchange in a networked organization, *Journal of the American Society for Information Science*, 46(12), 1101–14.

Henrich, J. (1999). Cultural transmission and the diffusion of innovations. Working Paper 99–020, University of Michigan Business School, Ann Arbor, MI.

Jang, S., Dai, S., Sung, S. (2005). The pattern and externality effect of diffusion of mobile telecommunications: The case of the OECD and Taiwan. *Information Economics and Policy*, *17* (2005), 133–148.

Johnson, S. (2001). *Emergence*. New York: Simon & Schuster.

Levin, S., Levin, S., & Meisel, J. (1987). A dynamic analysis of the adoption of a new technology: The case of optical scanners. *Review of Economics and Statistics*, 69, 12–17.

Macleod, C. (1992). Strategies for innovation: The diffusion of new technology in nineteenth century British industry. *Economic History Review*, 45, 285–307.

Macy, M. W., & Willer, R. (2002). From factors to actors: Computational sociology and agent-based modeling. *Annual Review of Sociology*, 28, 143–166.

Mahler, A. & Rogers, E. M. (1999). The diffusion of interactive communication innovations and the critical mass: The adoption of telecommunications services by German banks. *Telecommunications Policy*, *23*(10/11), 719–740.

Markus, M. L. (1987). Toward a "critical mass" theory of interactive media: Universal access,interdependence and diffusion. *Communication Research*, *15*(5), 491–511.

Markus, M. L. (1990). Toward a "critical mass" theory of interactive media. In J. Fulk & C. Steinfield (Eds), *Organizations and communication technology*, (pp. 194–217). Newbury Park, CA: Sage.

Midgley, D., Morrison, P. & Roberts, J. (1992). The effect of network structure in industrial diffusion processes. *Research Policy*, *21*, 533–552.

Monge, P. R. & Contractor, N. S. (2001). Emergence of communication networks. In F. M. Jablin & L. L. Putman (Eds), *The new handbook of organizational communication: Advances in theory, research, and methods* (pp. 440–502). Thousand Oaks, CA: Sage.

Monge, P. R. & Contractor, N. S. (2003). *Theories of communication networks*. New York: Oxford University Press.

Monge, P. R. & Eisenberg, E. M. (1987). Emergent communication networks. In F. M. Jablin, L. L. Putnam, K. H. Roberts & L. W. Porter (Eds), *Handbook of organizational communication: An interdisciplinary perspective* (pp. 304–342). Newbury Park, CA: Sage.

Moore, G. C., & Benbasat, I. (1991). Development of an instrument to measure the perceptions of adopting an information technology innovation. *Information Systems Research, 2*(3), 192–222.

Newman, M. E. J. (2002). The structure and function of networks. *Computer Physics Communications, 147*, 40–45.

Nyblom, J., Borgatti, S., Roslakka, J., & Salo, M. A (2003). Statistical analysis of network data-an application to diffusion of innovation. *Social Networks, 25*(2), 175–195.

Palmer, D. A., Jennings, P. D., & Zhou, X. (1993). Late adoption of the multidivisional form by large U.S. corporations: Institutional, political, and economic accounts. *Administrative Science Quarterly, 38*(1), 100–131.

Poole, M. S., & DeSanctis, G. (1990). Understanding the use of group decision support systems: The theory of adaptive structuration. In C. Steinfield & J. Fulk, J (Eds.), *Organizations and communication technology* (pp. 175–195). Newbury Park, CA: Sage.

Powell, W. W., Koput, K. W., & Smith-Doerr, L. (1996). Interorganizational collaboration and the locus of innovation: Networks of learning in biotechnology. *Administrative Science Quarterly, 41*(1), 116–145.

Quirmbach, H. (1986). The diffusion of new technology and the market for information. *Rand Journal of Economics*, 17, 33–47.

Reinganum, J. (1981). Market structure and the diffusion of new technology. *Bell Journal of Economics 12*, 618–624.

Rice, R. E. (1994). Network analysis and computer-mediated communication systems. In S. Wasserman & J. Galaskiewicz (Eds.), *Advances in social network analysis: Research in the social and behavioral sciences* (pp. 167–203). Newbury Park, CA: Sage.

Rice, R.E. & Danowski, J. A. (1991). Comparing comments and semantic networks about voicemail. *Proceedings of the American Society for Information Science, 28*, 134–138.

Rice, R., & Danowski, J. (1993). Is it really just like a fancy answering machine? Comparing semantic networks of different types of voice mail users. *Journal of Business Communication, 30*, 369–397.

Richards, W. D. & Rice, R. E. (1981) The Negopy network analysis program. *Social Networks, 3*, 215–223.

Rogers, E. M. (1962). *Diffusion of innovations*. New York: Free Press.

Rogers, E. M. (1971). *Communication of innovations*. New York: Free Press.

Rogers, E. M. (1983). *Diffusion of innovations, third edition*. New York: Free Press.

Rogers, E. M. (1995). *Diffusion of innovations, fourth edition*. New York: Free Press.

Rogers, E. M. (2003). *Diffusion of innovations, fifth edition*. New York: Free Press.

Rogers, E. M. & Kincaid, D. L. (1981). *Communication networks: Toward a new paradigm for research*. New York: Free Press.

Rosenkopf, L., & Abrahamson, E. (1999) Modeling reputational and informational influences in threshold models of bandwagon innovation diffusion. *Computational & Mathematical Organization Theory, 5*(4), 361–384.

Scott, J. (1991). Networks of corporate power. *Annual Review of Sociology, 17*, 181–203.

Scott, J. (2000). *Social network analysis: A handbook, second edition*. Thousand Oaks, CA: Sage.

Seary, A., Riopelle, K., & Richards, W. (2008). MultiNet/Negopy [computer program]. Burnaby, B.C.: Simon Frazier University.

Shah, P. P. (2000). Network destruction: The structural implications of downsizing. *Academy of Management Journal*, *43*(1), 101–112.

Snijders, T. A. B. (2001). The statistical evaluation of social network dynamics. In M. E. Sobel & M. P. Becker (Eds), *Sociological methodology* (pp. 361–395). Boston and London: Basil Blackwell.

Strogatz, S. (2003). *Sync*. New York: Hyperion.

Tarde, G. (1903). *The laws of imitation*. New York: Henry Holt.

Torenvlied, R., & Velner, G. (1998). Informal networks and resistance to organizational change: The introduction of quality standards in a transport company. *Computational & Mathematical Organization Theory*, 4(2), 165–189.

Utterbeck, J. (1994). *Mastering the dynamics of innovation*. Cambridge, MA: Harvard Business School Press.

Valente, T. W. (1995). *Network models of the diffusion of innovations*. Cresskill, NJ: Hampton.

Valente, T. W. (1996). Social network thresholds in the diffusion of innovations. *Social Networks*, *18*, 69–89.

Valente, T. W. (2005). Network models and methods for studying the diffusion of innovations. In P. Carrington, S. Wasserman & J. Scott (Eds), *Recent advances in network analysis* (pp. 98–116). Cambridge, UK: Cambridge University Press.

Valente, T. W. & Davis, R. L. (1999). Accelerating the diffusion of innovations using opinion leaders. *The Annals of the American Academy of Political and Social Science*, 566(1), 55–67.

von Seggern, D. (2007). *Standard curves and surfaces with mathematics, 2nd ed*. Boca Raton, FL: CRC.

Waarts, E., van Everdingen, Y. M., & Hillegersberg, J. (2002). The dynamics of factors affecting the adoption of innovations. *Journal of Product Innovation Management*, *19*(6), 412–423.

Walden, E., & Brown, G. (2002). Information cascades in the adoption of new technology. In F. Miralles, J. Valor, & J. DeGross (Eds), *Proceedings of the Twenty-Third International Conference on Information Systems* (pp. 435–443). Barcelona, Spain, December 15–18, 2002.

Watts, D. (1999). *Small worlds: The dynamics of networks between order and randomness*. Princeton, NJ: Princeton University Press.

Watts, D. (2003). *Six degrees: The science of a connected age*. New York: Norton.

Wellman, B. (1983). Network analysis: Some basic principles. In R. Collins (Ed), *Sociological theory* (pp. 155–200). San Francisco: Jossey-Bass.

Westphal, J. D., Seidel, M. L., &Steward, K. J. (2001) Second-order imitation: Uncovering latent effects of board network ties. *Administrative Science Quarterly*, 46(4), 717–747.

· 7 ·

Network Diffusion of Two Competing Ideas

An Application of Agent-Based Modeling

FRANK TUTZAUER, KYOUNGHEE HAZEL KWON & BENJAMIN ELBIRT

Introduction

Agent-based modeling is a simulation technique in which agents, governed by local rules of behavior, interact in ways that often generate and, at the same time, constrain global behavior. In this chapter, we present an agent-based model of the diffusion innovations. Rather than the classical case in which a single innovation diffuses throughout a population over time, we consider the case of two competing innovations. We also examine three different rules of adoption: whether agents adopt innovations based on the behavior of influential neighbors, whether they adopt innovations of structurally equivalent others, or whether they give more weight to longtime adopters compared with recent adopters. Results suggest that centrality is an important predictor of the diffusion of innovations.

Researchers interested in the diffusion of innovations have investigated various factors that determine an individual's adoption decision. The factors range widely, from characteristics of the innovation—such as compatibility, relative advantage, complexity, and trialability—to characteristics of the adopters—such as socio-demographic factors and psychological and personality characteristics—to external or environmental conditions—such as marketing, governmental regulation, and cultural climate (Rogers, 1995; Wejnert, 2002). Interpersonal communication, in particular, is one of the most widely discussed mechanisms of adoption and is found in a variety of classical diffusion models such as a Bass' (1969/2004) imitation model, Granovetter's threshold model, the two-step flow hypothesis (Katz, 1957; Weimann, 1982), and social network models (Valente & Davis, 1999; Valente, 2005).

The importance of interpersonal influence in the diffusion of innovations leads one naturally to social network theory and analysis. The social network perspective highlights the interdependent nature of social relationships, the impact of interpersonal ties on agent behaviors, and the over-time changes in a community, an organization, or a society. Networks are primarily communication patterns, formed from informal talks and advice from friends, families, and neighbors, and technical, professional, or formal communication among members of a community or organization. Social networks have been understood as channels through which individuals influence the spread of innovation (DeBresson & Amesse, 1991).

The current study aligns with those models that pay special attention to interpersonal influence on diffusion processes. We borrow Valente's (2005) notion of *Personal Network Exposure*, or PNE, which is the degree to which an individual witnesses others' adoption behaviors within his or her personal network. While the likelihood of an individual's adoption is determined by the level of PNE, computation of PNE can be varied depending on which particular property is understood as a major source of influence. We consider three different factors—opinion leadership, structural equivalence, and time since latest adoption—that may be exposed to potential adopters through their personal network.

Although the network model of diffusion is a useful approach to understanding social influences of adoption behaviors, it nonetheless has practical drawbacks. Network models require data not only about the longitudinal change of innovation penetration within a system but also on all pairs of individuals in the network. For example, if a system has 100 members, there are 4,950 pairs of individuals that either are or are not connected, and the researcher also has to keep track of who has adopted the innovation and when the adoption occurred. The amount of empirical data collected can be quite large.

An alternative to empirical data is simulated data. In this chapter, we use agent-based modeling (ABM) as an alternative means of exploring network effects on the diffusion process. ABM is a simulation technique that highlights the complex nature of social systems. As a web of interconnected actors, the social system changes over time, not by a fixed set of causal variables but by (possibly nonlinear) interactions among actors within the system. Although actors may follow a simple set of specified rules, the consequential change on the macro-level can be quite complex. Using ABM, a researcher can simulate agents' behaviors under these simple rules and observe how the system undergoes changes as a whole, driven by the members' micro-behaviors. Because it is not always easy to keep track of empirical micro-interactions among actors, an agent-based simulation is a complementary approach to the data-driven causal model of diffusion (Garcia, 2005).

Accordingly, the current chapter makes several contributions: First, we compare PNE measures based on different sources of social influence. Second, we examine the diffusion of not just one innovation, but two, a situation more realistic than the classic case of a single innovation. Finally, this chapter shows how the application of ABM to the network models of diffusion studies can produce fruitful avenues of inquiry.

The structure of the chapter is as follows: First, we review previous network models of diffusion and introduce the key adoption mechanism, namely, personal network exposure. Then we give a brief description and review of agent-based modeling. Next we examine three different means of social influence and show how they can be incorporated into the measurement of personal network exposure. Finally, we present the details of an agent-based simulation, report the results of the simulation, and offer some concluding thoughts.

Network Models of Diffusion

Basic Network Concepts

A *network* consists of a set of *nodes*, numbered 1 through N, and a set of *ties* that link nodes together. Because we are looking at the diffusion of innovations among individuals—individuals who are represented by the nodes of a network, but are simulated by agents in an agent-based model—we will use the terms *individual*, *node*, and *agent* interchangeably. The ties are the basic connections between pairs of individuals, and these ties may or may not be directional. In an *undirected* network, a tie between nodes i and j is reciprocated by a tie from j to i. In a *directed* network, ties are not always reciprocated. In either event, the *density* of the network is the number of ties that it has divided by the number of possible ties.

If there is a tie from i to j, then i and j are said to be *adjacent* and j is a *neighbor* of i. We define the $N \times N$ *adjacency matrix* $A = [a_{ij}]$ by setting a_{ij}, the matrix entry in the ith row and jth column of A, to be equal to 1 if there is a tie from i to j, and 0 otherwise. A node's *personal network* consists of the node's neighbors along with the ties amongst them. The number of neighbors a node has is the node's *degree*, which is given by the appropriate row sum of the adjacency matrix. In other words, the degree of node i is given by Σa_{ij}, where the sum is to be taken across all j.

One of the most important nodal properties is how central the node is to the network. There are a variety of extant centrality measures. In this article, we use four: *degree centrality*, which is simply the node's degree; *closeness centrality*, which is based on the distance between the focal node and all other nodes in the network—the closer the node is to other nodes, the more central it is; *betweenness centrality*, which indicates how often a focal node is between other nodes in the sense of being able to shut off communication from one node to another; and *eigenvector centrality*, which measures a node's centrality on the basis of whether or not the node is connected to other nodes of high centrality. A review of all of these centrality measures can be found in Wasserman and Faust (1994).

Networks and Diffusion

Since the pioneering studies of communication networks by Coleman, Katz, and Menzel (1957) and Rogers and Beal (1958), decades of diffusion research have shown that diffusion is a social process susceptible to interpersonal influence. That interpersonal influence is an essential component in the diffusion process is supported by many well-known diffusion models. For example, Bass (1969/2004) proposed that the probability of adoption depends on the number of previous adopters, modified by two parameters: an innovation parameter governing the number of innovators (adopters at time 0) and an imitation parameter governing the extent to which individuals imitate others in the system. Another widely applied model, Granovetter's threshold model (1987), also assumes that individuals choose to adopt a behavior when they perceive a certain rate of adoption among other members of a community.

Both of these models highlight the importance of interpersonal influence, but, strictly speaking, neither is a network model. Granovetter and Bass assume that adoption decisions are made based on what is happening in the population as a whole. A more realistic assumption is that interpersonal influence operates by means of those to whom one is close in a social, physical, or structural sense.

Valente and Davis (1999) embraced this more realistic assumption by explicitly using social networks. Their network model highlights the importance of our immediate social surroundings, so-called personal networks. According to Valente and Davis, rather than the adoption rate in the entire population, what is most important is the adoption rate in one's personal network; an individual adopts the innovation only if the proportion of adopters in his or her personal network exceeds some threshold. Hence, it is the adoption decisions of one's immediate interpersonal contacts that are important rather than those of more socially distant individuals.

Valente (2005) generalized this concept, calling it *Personal Network Exposure*. The underlying assumption is that individuals make decisions based, not on perfect knowledge about the whole community, but on partial knowledge formed through interactions within a personal network. If the level of personal network exposure is greater than a specified threshold, then the individual adopts the innovation. Different members in an agent's personal network may have differing levels of interpersonal influence on the agent, and these levels of influence are captured in a weighting matrix, $\mathbf{W} = [w_{ij}]$, where w_{ij} is the influence that agent j has on agent i. The particular source of influence might be structural, such as an agent's centrality in the network, or it might be a non-network feature, such as whether an agent is an early or a late adopter. A given agent's influence on others might be constant, such as the previously mentioned example of centrality, in which case w_{ij} would be the same across all agents i, or it might be a feature such as structural equivalence that varies across agents, in which case the w_{ij} are not constant across i. To compute agent i's PNE, one adds the weights of those neighbors of i who have adopted the innovation and then divides by the total possible. In particular, agent i's personal network exposure, E_i, can be computed as

$$E_i = \frac{\sum_j w_{ij} y_j}{\sum_j w_{ij}} \quad \text{[Equation 1]}$$

where y_j is equal to 1 if agent j has adopted the innovation and is equal to 0 otherwise, and where the sums are to be taken across all nodes j in i's personal network. The numerator of the above expression sums the weights of those neighbors of i who have adopted the innovation; the denominator, which is merely the sum of the weights of all individuals in agent i's personal network, serves to put the measure on a 0–1 scale. In the case where the weights are 1 or 0, depending on whether or not i and j are adjacent (i.e., where the weighting matrix is equal to the adjacency matrix), then E_i is simply the proportion of i's neighbors that have adopted the innovation, as in Valente and Davis (1999).

Competing Innovations

Although much diffusion literature highlights the diffusion processes of a single innovation, in reality it is common to observe situations in which two or more innovations compete with each other to attract adopters within the same market. For example, Häagen-Dazs and Ben & Jerry's compete within the same ice cream market. As another example, consider that two contrasting ideas are often spread simultaneously about a particular innovation. For example, when a new movie is released, positive and negative comments are spread, and our decision to watch is often affected by the level of exposure to each type of comment. Our simulation model considers the situation where two competing innovations or ideas are introduced, rather than just a single innovation as in the classic case.

The model is exemplified by the following scenario: Consider a group of agents who have not yet purchased a mobile phone. They make up a social network in which each agent is connected to at least one agent, thus nobody is isolated. There are two mobile carriers offering the latest high-tech phone options—let's suppose Verizon and AT&T—and agents will choose one over the other when they perceive higher popularity: if an individual perceives higher popularity of Verizon, the individual will adopt Verizon over AT&T and vice versa. If a person does not perceive predominance of either carrier, the person delays the adoption decision. In other words, agents follow the majority choice of previous adopters in the personal network.

Of course, the majority choice does not have to be a strict count of adopters for each innovation. This "majority" is modified by the influence that each of the agents has on others, that is, by the weights in the weighting matrix **W**. Moreover, one's observation cannot be perfect because an individual does not have a bird's-eye view of the whole community. Accordingly, agents in our model decide whether to adopt, and if so which of two innovations to adopt, based on their immediate environment that determines the level of PNE.

Rather than having an overall threshold, as in Valente and Davis (1999) and Valente (2005), we will let the innovations compete against each other, computing a PNE for each innovation. An agent will then adopt the innovation that has the higher PNE value. Stated in terms of our example, an agent will adopt Verizon if and only if the agent's PNE for Verizon is larger than the agent's PNE for AT&T. Similarly, an agent adopts AT&T if and only if the AT&T PNE is larger than that for Verizon.

Agent-Based Modeling (ABM)

Agent-based modeling is a simulation technique whereby rules of local interaction among agents give rise to global patterns. In other words, it is a methodology for studying "how complex outcomes flow from simple schemata and depend on the way in which agents are interconnected" (Anderson, 1999, p. 220). Macy and Willer (2002) argued that ABM is different from other types of simulation. According to Macy and Willer, two types of social simulation predated ABM: macrosimulation and microsimulation. Macrosimulation uses computers to simulate the impact of systemic factors on a population distribution. Typically, a model is composed of a set of differential equations. Microsimulation, on the other hand, like ABM, is a bottom-up approach that uses individuals as units of analysis. Given that the focus is on individuals, microsimulation and ABM are similar. However, two important aspects distinguish ABM from the preexisting simulation methods. First, while the primary goal of macrosimulation and microsimulation is "empirically based macro-level forecasting," ABM is a kind of "thought experiment" that pursues theoretical development rather than prediction (Macy & Willer, 2002, pp. 145–147). Second, unlike microsimulation, in which actors are socially isolated, ABM highlights local interactions among agents who are assumed to be autonomous, adaptive, and interdependent.

A classic use of ABM is Reynolds' (1987) modeling of flocks of birds. In his model, birds follow basic rules that define their behaviors in relation to other birds. Despite the simplicity of rules that determine the movement of individual birds, the flock as a whole is an extremely complex system that cannot be easily explained by a top-down account of global behavior. As in the Reynolds' model, the main goal of ABM is to observe how micro-behaviors result in complex macro-phenomena. Local interactions among individual actors are performed based on simple shared rules. However, the outcome of the local interactions as a whole is often nonlinear, which is difficult to capture with standard causal analyses of a set of factors. The advantage of ABM is realized when the researcher is interested in a system's emergent properties such as cultural convergence, segregation, social influence, or imitation (e.g., Axelrod, 1997; Carley, 1991). The diffusion of innovations is a fruitful area for ABM, particularly when one is interested in the spread of information or social influence through social networks (e.g., Bonabeau, 2002; Delre, Jager, & Janssen, 2007; Guardiola, Díaz-Guilera, Pérez, Arenas & Llas, 2002; Valente, 1995; Young, 2006).

As Garcia (2005) has noted, ABM is particularly useful when "both macro- and microlevels of analyses are of interest" and when "social systems can easily be described by 'what-if' scenarios but not by differential equations" (p. 383). In this chapter, we use ABM for three separate reasons: First, although diffusion is a macro-level phenomenon, the dynamics of diffusion are not effectively captured without observing micro-level adoptions. Second, we want to delineate different patterns of diffusion according to different "what-if" scenarios that capture different ways that personal network exposure affects the individual's decision to adopt. These variations in decision rules cannot be measured or operationalized as an analytic attribute. Finally, we conceive of diffusion in social networks as a competitive market of innovations. The competition between innovations implies that the two different innovations are coevolving within the same environment.

Three Approaches to Personal Network Exposure

Assume that we have two competing innovations, A and B. We consider three different weighting factors for PNE: opinion leadership, structural equivalence, and time since latest adoption. Because we have two competing innovations, we will calculate two separate PNE measures for each weighting possibility: A_i, agent *i*'s PNE for innovation A, and B_i, the equivalent PNE for innovation B. Because we are not using an absolute threshold, but, rather, are comparing the PNEs to each other, there is no reason to normalize to a 0–1 scale, and hence we can simplify the equations by not dividing by an agent's sum of weights. In other words, our PNEs will be given by

$$A_i = \sum_j w_{ij} x_j \qquad \text{[Equation 2]}$$

and

$$B_i = \sum_j w_{ij} y_j, \qquad \text{[Equation 3]}$$

where x_j is equal to 1 if j is one of i's neighbors who has adopted innovation A, and is equal to 0 otherwise (y_j is defined similarly for innovation B), where w_{ij} is the weight (influence) that j has on i, and where the sums are to be taken across all individuals in i's personal network. These weights will be different depending on whether we are considering opinion leadership, structural equivalence, or time since latest adoption, and their computation is discussed next.

Opinion Leadership

In network diffusion studies, opinion leadership has generally been based on sociometric position. For example, a name nomination approach has been used to identify those who are contacted frequently for advice or information within a community. These highly nominated individuals are considered highly influential (Becker, 1970; Coleman et al., 1957; Valente & Davis, 1999). Weiman (1991) found further support for the positional measure of opinion leadership by showing a high correlation between centralities in communication networks and a personality-based assessment of influence ability. Accordingly, we assume that the social network may be composed of peer nominations, and we use degree centrality as the indicator of opinion leadership. Then, the weights in the PNE measures are simply the degrees of the nodes:

$$w_{ij} = \sum_k a_{jk} \qquad \text{[Equation 4]}$$

Structural Equivalence

Valente (1995, 2005) suggests structural equivalence as one means for interpersonal influence. Structural equivalence is the degree to which "two individuals occupy the same position in a social system" (Valente, 1995, p. 56). The idea is that the greater the structural similarity between a current adopter and the potential adopter, the more likely the potential adopter is to actually adopt the innovation. Burt (1987) shows that weighting the network effects by structural equivalence scores predicted an individual's time of adoption better than considering simply direct ties.

The structural equivalence between agents i and j can be computed with a Euclidean distance formula utilizing the entries of the adjacency matrix:

$$SE_{ij} = SE_{ji} = \sqrt{\sum_k [(a_{ik} - a_{jk})^2 + (a_{ki} - a_{kj})^2]} \qquad \text{[Equation 5]}$$

Note that in the above formula, if agent i and agent j have ties to exactly the same other agents, then inside each of the above sets of parentheses the two terms will either be both equal to 1 or both equal to 0; hence, the sum will also be 0. Consequently, the lower SE_{ij}, the more structurally equivalent the two agents are. We wish to have our PNEs so that larger weights (not smaller) correspond to greater influence. Structural equivalence, as defined above, is in the "wrong direction," with larger scores representing less structural similarity and, therefore, less influence. We can make the scores run in the correct direction by subtracting from the maximum. In other words, for structural equivalence we set the weights as follows:

$$w_{ij} = M - SE_{ij} \qquad \text{[Equation 6]}$$

where $M = \max\{SE_{ij}\}$ is the maximum structural equivalence score amongst all pairs of nodes in the network. The resulting weights are thus similarities; the greater the similarity, the greater the weight, and thus the more influence j exerts upon i.

Time Since Latest Adoption

One important variable that should be considered is the time since adoption. An agent is more likely to be influenced by an individual who adopted the innovation quite some time ago as opposed to an individual who only recently adopted the innovation. For example, Watts and Dodds (2007) simulated the degree of influence in diffusion of public opinions. Although they used the term *opinion leaders*, their consideration of the influential individuals was not based on a sociometric definition of opinion leaders (as ours is, above). Instead, they examined the impact of initiators or early adopters as opposed to late adopters. Accordingly, another possible weighting factor in measuring PNE is the length of time the innovation has been adopted by an individual.

As will be seen in the Method section, we consider two separate cases regarding adoption time. First, we consider the classic case that distinguishes between adopters and nonadopters. Individuals who become adopters remain adopters for all time. Because we are dealing with two innovations, however, it makes sense to allow individuals to switch between innovations—in other words we allow them to change their minds. We will therefore take as our weights the time since latest adoption (rather than the time since initial adoption).

Our simulations proceed in discrete steps. In other words, we start at time 0 and continue thence to time 1, time 2, etc. Each time step is called an *iteration*. Let t be the current iteration in the simulation and suppose that at iteration t agent j is a current adopter of innovation A. Let $t(j)$ be the most recent iteration in which agent j was *not* an adopter of innovation A. Then the appropriate weighting score to use in the computation of A_i is:

$$w_{ij} = \frac{t - t(j)}{t}. \qquad \text{[Equation 7]}$$

The above expression is the proportion of time that agent j's most recent adoption has been innovation A. Of course, the weights can be defined similarly for those agents that are current adopters of innovation B.

With Equation 7 we have come to the end of our theoretical exposition. Using these various weights for the computation of PNE, we conducted a simulation using three empirical networks. In brief, for each network we chose a pair of innovators (one for A, one for B), and we allowed the innovations to propagate throughout the network according to the three different weighting rules above. Details of the simulation can be found in the next section.

Method

Empirical Networks

The simulation used three types of real life social networks: a Facebook friendship network, an advice network, and a network of jazz bands. The Facebook network consists of a user's 112 friends. This network comes from a larger data set from one of the author's (Kwon) dissertation project. Although the average size of personal networks in Facebook tends to be larger, often reaching over 300 individuals, we intentionally chose a smaller size to lessen the computational burden in the simulation. The ties are binary, with a tie being placed between *i* and *j* if *i* and *j* are friends. In Facebook, two individuals become friends only when there is mutual agreement. Accordingly, this network is undirected. It has 1,481 ties and a density of 0.12.

The advice network consists of 77 individuals in a research team for a manufacturing company (this network comes from the research of Cross and Parker [2004] and is available at http://toreopsahl.com/datasets/). In its original form, this was a directed network with a tie from *i* to *j* indicating that *i* sought *j*'s advice. In our simulation, we used a symmetrized version by adding reciprocal ties wherever an unreciprocated tie was found. Although there is nothing in our theoretical analysis that requires symmetric networks, we felt it best not to introduce this particular confounding condition, keeping all of our networks undirected. This way, any difference between the advice network and other networks will not be due to its status as a directed network. In the symmetrized version of the network, there are 2,682 reciprocated ties for a density of 0.46.

Finally, the jazz network consists of 198 bands that performed between 1912 and 1940, with a tie placed between two bands if they shared a common musician. There are 5,484 such ties, giving the network a density of 0.15. (The jazz network was first discussed by Gleiser and Danon [2003], and the data set can be found at http://deim.urv.cat/~aarenas/data/welcome.htm.)

Design

Our simulation consists of a large number of *runs*. We used a 3 × 3 × 2 design (Network × PNE × Decision Type) to set the parameters of each run. In other words, each run appears in exactly one cell of the 3 × 3 × 2 design. The three different networks were the Facebook, advice, and jazz networks discussed above. The three different PNEs were those where the weights were based on our three influence factors: opinion leadership, structural equivalence, or time since latest adoption. The two decision types were fixed and floating. In the fixed-decision condition, once an agent adopted an innovation, whether A or B, that agent stayed with that innovation until the end of the run. In the floating-decision condition, we allowed the agents to change their minds. For example, an agent adopting innovation A will change to innovation B if the agent's PNE for B becomes greater than the PNE for A, a result that might happen as the other agents in the first agent's personal network make their own decisions about which innovation, if either, to adopt.

Procedure

At each iteration, each agent is in one of three states, A, B, or N, depending on whether the agent is currently an adopter of innovation A, innovation B, or neither and thus is neutral. At the beginning of each run, we choose a pair of agents to act as *seeds*. One of the two seeds is placed in state A and one in state B. All of the other agents begin the run in state N. Then, in each iteration, we use Equations 4, 6, or 7, depending on PNE condition, to compute weights for each agent, and we use Equations 2 and 3 to compute the PNEs A_i and B_i for each agent (in the floating-decision condition) or for each agent currently in state N (for the fixed-decision condition). Then, for every agent that is eligible to change states (i.e., all agents in the floating-decision condition, but only those agents in state N for the fixed-decision condition) we use the following updating rule: in the next iteration, place agent *i* in state A if $A_i > B_i$, in state B if $A_i < B_i$, and in state N if $A_i = B_i$. The iterations continue in this fashion until the agents converge to a steady state, in which no agent ever changes states again, or until the maximum number of iterations (100) is reached.

At the end of a run, we record the number of iterations in the run and the saturation level of each innovation (i.e., the proportion of agents in state A at the end of the run, and similarly for innovation B), and we choose a new pair of seeds for the next run. We conduct a separate run for each possible pair of seeds, thus the number of runs in each network condition varies on the basis of network size. In particular, there were 6,216 runs in each of the Facebook cells of the design, 2,926 in the advice cells, and 19,503 in the jazz musician cells. In all, across the 18 cells of the design, we conducted 171,870 runs.

Independent and Dependent Variables

We are interested in the structural properties of seed agents in terms of how their innovations spread through the network. Consequently, we chose as our independent variables degree centrality, closeness centrality, betweenness centrality, and eigenvector centrality. To compute the centralities, we used UCINET 6 (Borgatti, Everett & Freeman, 2002).

Regarding dependent variables, we gave a seed credit for a *win* if the innovation that the seed began with at the outset of the run was adopted by a plurality of the agents in the network at the end of the run. In a network of size *N*, in each cell of the design each agent is used as a seed *N*–1 times (once with each of the other *N*–1 agents), thus the number of wins for an agent in a given cell of the design can vary from 0 to *N*–1. If we look at the saturation level of a seed's innovation in each of the seed's wins, we can compute the mean saturation level in runs where the agent won. We can also compute a similar mean for time to convergence, or what we call *saturation time*. These three variables—win count, mean saturation level, and mean saturation time—serve as our main dependent variables.

Results

Centrality

As mentioned previously, we used various centrality measures to assess each seed's positional properties. Within each of the three networks, the four measures of centrality were significantly correlated, a result to be expected based on previous research (Valente, Coronge, Lakon & Costenbader, 2008). We obtained large correlations amongst degree, closeness, and eigenvector centrality, ranging from a low of 0.78 to a high of 0.98. The correlations between each of these three measures and betweenness centrality were smaller, ranging from a low of 0.26 to a high of 0.96.

After computing centralities, we correlated the scores with the diffusion outcome variables: mean saturation percentage (SP), mean saturation time (ST), and win count (WC). Tables 1–3 summarize the results. The most outstanding finding is that win counts show remarkably high correlations with most of the centrality scores; the more central the seed, the more wins obtained by that seed, regardless of network, PNE, or decision type. The only exceptions to this result were the two nonsignificant correlations with betweenness in the Facebook network: the structural equivalence version of PNE with fixed decisions and the adoption time version of PNE with floating decisions. In all other cells of the design, an innovation diffused by a central seed was very likely to dominate the market share. In general, the betweenness scores were moderately correlated with the win counts (ranging from $r = 0.21$ to $r = 0.58$), while the correlations between win counts and the rest of centralities scores were very strong (ranging from $r = 0.76$ to $r = 0.97$).

The results regarding saturation percentage varied depending on which PNE model was of interest and which network was examined. With the Facebook network, eigenvector centrality was negatively associated, and closeness centrality was positively associated, with saturation percentage when using the structural equivalence PNE in the fixed and floating decision-types, respectively. On the other hand, betweennness centrality was positively associated with saturation percentage in all of the other Facebook cells. In the advice networks, all of the centrality measures were associated with saturation percentage for two of the fixed-decision models, namely, when using an adoption time PNE or when using an opinion leadership PNE. None of the centrality measures predicted saturation percentage in the other advice network cells. In the jazz network, the results are mixed, with different centralities predicting saturation percentage to different extents in the various decision-type and PNE conditions.

Likewise, the effect of a seed's network position on saturation time seems to vary from model to model. The most consistent results are for the advice networks where all correlations save one (fixed/adoption time) were significant. In the other networks, the results were more varied. In all cases, significant coefficients were negative, indicating that the more central the seed, the less time it took to reach convergence.

Mean Comparisons

Table 4 shows the means and standard deviations of each of the dependent variables in each of the 18 cells of the design. Several consistent patterns can be observed. First of all, means for win counts were remarkably constant across different PNEs and decision types. There are big differences between networks, but this is an artifact of network size. In the smallest network—the advice network—each node is used as a seed 76 times, whereas in the largest network—the jazz network—each node is used as a seed 197 times. Naturally, there is more opportunity for a node to accumulate wins in the jazz network than in the advice network, with the Facebook network being intermediate. Nonetheless, within each network, the win counts are very close to each other. Neither PNE nor decision-type seems to affect the means.

Second, in general, the mean saturation percentage is high across all models, ranging from 70% to 99% with small standard deviations. This result indicates that regardless of which agent initiates the diffusion, one innovation is likely to win by a landslide against the competing one. The saturation percentages for fixed decisions seem to be somewhat smaller than for floating decisions. The means for floating decisions, in particular, suggest that we may be hitting a ceiling, and this ceiling effect probably attenuated the correlations we reported early, making patterns difficult to discern for mean saturation percentage.

With respect to saturation time, the fixed decisions once again are smaller than for the floating decisions, a situation that mirrors the results for saturation percentage. Naturally, if agents are not allowed to change their minds, then it is less likely that a single innovation totally takes over the network, and, moreover, it will take less time for the agents to come to a steady state. We can also see from Table 4 that opinion leadership results in quicker convergence for fixed decisions than does time since latest adoption.

Third, the mean scores among PNEs suggest that, in the fixed decision condition, leadership influence may be the most efficient both in terms of saturation percentage and time. Regarding the time spent to reach saturation, leadership and structural equivalence took equal time periods for saturation. However, the leadership-based PNE shows the highest saturation level across all networks.

PNE Correlations

Finally, within each cell of the design, we correlated the mean saturation percentages with each other by considering in a pair-wise fashion each of the three PNE weighting schemes. As Table 5 shows, the correlations are in general high, suggesting that, with respect to saturation percentage at least, the three PNE weighting schemes are largely interchangeable.

Discussion

The current study applied agent-based modeling to simulate diffusion processes of two competing innovations. Our model was based on the simple rule that individuals will follow the (appropriately weighted) majority choice when facing competing innovations, one of which an individual might adopt. The specific form of the model was based on Valente's (2005) personal network exposure, PNE. We investigated three different "what-if" scenarios embodied by three differently weighted versions of PNE—opinion leadership, structural equivalence, and adoption time—along with two decision-making conditions and three different real social networks.

Our findings support the important role of an innovator's network position, particularly when it comes to the chances of winning the competition. The winning frequencies were highly correlated with the centralities of the seed of the particular innovation. In our study, a seed's centrality was especially important because in most of the cases the winner dominated the market, with a minimum saturation of 70%, and usually higher. This finding suggests that it should be a strategic matter of choice "where to begin" as well as "how to spread." Subsequently, understanding positional properties of the potential innovator is an important prerequisite for success in a competitive market of ideas or innovations.

One emergent question related to the seed centralities is the influence of the discrepancy between the competing seeds' centralities on diffusion outcomes. We looked at each seed's centrality in isolation, but despite the pattern of correlations we obtained, it may be that it is not the seed's absolute centrality that is important, but rather the seed's centrality vis-à-vis the centrality of the seed beginning with the other innovation. For example, if innovation A's seed has a high centrality while innovation B's seed has a low centrality, the diffusion outcome might be different than when both seeds have equally high centralities. We leave answering this question to future research.

When mean saturation and saturation time are considered, our findings reveal inconsistencies across different models and different networks. The inconsistencies suggest that the saturation level and saturation time might be influenced more by both the specific network context where the competition occurs and the method by which the innovations diffuse than to seed centralities. One possibility is an interaction effect between network structures and the diffusion strategies. Network properties such as size, diameter, or centralization could interact with the different weights of PNE. Unfortunately, three networks are far too few to adequately explore this question, and thus we leave this issue to future research as well.

On the other hand, we observed that types of decision resulted in different mean saturation levels and different mean saturation times. Recall that we employed two decision types, one of which was fixed, in that once an agent adopts an innovation the agent never changes, and one of which was floating, in that agents could change from iteration to iteration as to which, if either, of the innovations they adopt. With floating decisions, saturation levels were higher and the saturation times were longer than with fixed decisions. Considering that it largely depends on the nature of innovation whether or not we could change our adoption decision easily, decision type can be understood as an indirect indicator of innovation characteristics. Within the same decision type, we found no drastic differences among different PNEs. High correlations among PNE models also reflect the interchangeability of the three weighting factors, which indicate that what mattered more in our simulation seemed to be the decision types, or the characteristics of innovation, than the factors that influenced the PNE level.

However, this result should not lead to the simple conclusion that network effects are trivial in the diffusion process. Rather, our results should lead to the pursuit of more advanced application of PNE weights. While our focus was to compare the different PNE weights, these weights could be at work simultaneously. It is probably unrealistic that all agents use the same weighting rules or that agents keep a single rule for all time. Therefore, future study can consider extending the network influence model by combining the three PNE factors in one scenario. The simultaneous consideration could be performed in various ways—for example, by adding the three weights together and applying the combined PNE to the whole agent population, by segregating populations into three subsets and applying different rules to the different subpopulations, or by taking into account certain agent attributes or times and accordingly applying different PNEs.

Our study has limitations that can be improved in future research. First of all, we applied a simple behavioral rule, a weighted majority. But, ABM is a kind of thought experiment, and it is up to a researcher's insight as to how to define the rules. Our rule can be considered as a simple kind of bandwagon effect, driven by dynamic sub-mechanisms (Abrahamson & Rosenkopf, 1997), and it can be improved by considering more refined rules for agent behaviors such as requiring larger margins than simple majority, or by considering second order neighborhoods (the personal networks of those in one's personal network). Second, we considered only binary, undirected social networks. Weighted networks or directed networks may present a quite different structural context and thus result in different diffusion patterns. Third, we did not consider the longitudinal pattern of diffusion processes. Observation of over-time processes is a common approach in diffusion studies, and it may well be explored with ABM. We did not include this aspect in the current study because of computational difficulties: Running and computing the results of each of the 18 cells of the design, at each iteration, required an overwhelming amount of time. The same computational burden also prevented us from modeling diffusion processes with multiple seeds. It is more realistic, however, to consider more than a single seed to begin the diffusion of each innovation. A next step would be to consider multiple innovators in the model.

This chapter proposed the utility of ABM in exploring how interpersonal influence on the micro-level affects the diffusion pattern on the macro-level. Diffusion researchers can take advantage of ABM especially when it is difficult to acquire empirical data. The network diffusion model of competing innovations, which we simulated in our study, is one good example for which data collection is so daunting that few empirical studies could have been conducted, even if such situations are commonly observed in our social life. ABM is a complementary and alternative methodology to the existing trend of diffusion research.

Table 1. Correlations between seed centralities and diffusion variables: Leadership PNE models.

Networks		Facebook				Advice				Jazz		
		SP	ST	WC		SP	ST	WC		SP	ST	WC
	DG	n.s.	n.s.	.97***	DG	n.s.	-.31**	.86***	DG	.21**	-.28**	.85***
Floating	CL	.24*	n.s.	.91***	CL	n.s.	-.29*	.76***	CL	.28**	-.14*	.86***
	BT	n.s.	-.34**	.22*	BT	n.s.	-.24*	.58***	BT	.22*	-.19*	.32***
	EV	n.s.	n.s.	.97***	EV	n.s.	-.30**	.93***	EV	n.s.	-.34***	.94***

Networks		Facebook				Advice				Jazz		
		SP	ST	WC		SP	ST	WC		SP	ST	WC
	DG	n.s.	n.s.	.97***	DG	.49***	-.55***	.86***	DG	.28**	n.s.	.86***
Fixed	CL	n.s.	-.30**	.92***	CL	.47***	-.61***	.76***	CL	.26**	n.s.	.89***
	BT	.19*	-.23*	.23*	BT	.44***	-.66***	.58***	BT	.26**	-.15*	.36***
	EV	n.s.	n.s.	.96***	EV	.45***	-.46***	.93***	EV	n.s.	.21*	.91***

Note. SP = mean saturation percentage; ST = mean saturation time; WC = win counts; Centralities: DG = degree; CL = closeness; BT = betweenness, EV = eigenvector; * $p < .05$, ** $p < .01$, *** $p < .001$, n.s. = not significant

Table 2. Correlations between seed centralities and diffusion variables: Structural equivalence PNE models.

Networks		Facebook				Advice				Jazz		
		SP	ST	WC		SP	ST	WC		SP	ST	WC
	DG	n.s.	-.25*	.92***	DG	n.s.	-.28*	.81***	DG	n.s.	-.21**	.77***
Floating	CL	n.s.	-.31***	.84***	CL	n.s.	-.28*	.71***	CL	n.s.	-.15*	.86***
	BT	.19*	-.26*	n.s.	BT	n.s.	-.29*	.51***	BT	n.s.	n.s.	.29***
	EV	n.s.	n.s.	.96***	EV	n.s.	-.24*	.90***	EV	n.s.	-.22**	.85***

Networks		Facebook				Advice				Jazz		
		SP	ST	WC		SP	ST	WC		SP	ST	WC
	DG	n.s.	n.s.	.94***	DG	n.s.	-.47***	.83***	DG	n.s.	n.s.	.80***
Fixed	CL	n.s.	-.30***	.91***	CL	n.s.	-.53***	.72***	CL	n.s.	n.s.	.88***
	BT	n.s.	-.24*	n.s.	BT	n.s.	-.56***	.53***	BT	n.s.	-.15*	.32***
	EV	-.28**	n.s.	.96***	EV	n.s.	-.40***	.90***	EV	n.s.	.19**	.86***

Note. SP = mean saturation percentage; ST = mean saturation time; WC = win counts; Centralities: DG = degree; CL = closeness; BT = betweenness; EV = eigenvector; * $p < .05$, ** $p < .01$, *** $p < .001$, n.s. = not significant.

Table 3. Correlations between seed centralities and diffusion variables: Adoption-time PNE models.

Networks		Facebook				Advice				Jazz		
		SP	ST	WC		SP	ST	WC		SP	ST	WC
	DG	n.s.	-.28**	.92***	DG	n.s.	-.52***	.85***	DG	.22**	-.36***	.79***
Floating	CL	n.s.	-.34***	.84***	CL	n.s.	-.50***	.75***	CL	.20**	-.21**	.86***
	BT	.25**	-.29**	n.s.	BT	n.s.	-.45***	.58***	BT	.23**	-.15*	.30***
	EV	n.s.	-.19*	.96***	EV	n.s.	-.48***	.92***	EV	n.s.	-.33***	.87***

Networks		Facebook				Advice				Jazz		
		SP	ST	WC		SP	ST	WC		SP	ST	WC
	DG	n.s.	n.s.	.95***	DG	.70***	n.s.	.86***	DG	.18*	n.s.	.81***
Fixed	CL	n.s.	-.29**	.92***	CL	.69***	n.s.	.76***	CL	n.s.	n.s.	.88***
	BT	.19*	-.21*	n.s.	BT	.65***	n.s.	.58***	BT	.22**	-.15*	.33***
	EV	n.s.	n.s.	.96***	EV	.65***	n.s.	.92***	EV	n.s.	n.s.	.88***

Note. SP = mean saturation percentage; ST = mean saturation time; WC = win counts; Centralities: DG = degree; CL = closeness; BT = between ness; EV = eigenvector; * $p < .05$, ** $p < .01$, *** $p < .001$, n.s. = not significant.

Table 4. Mean comparisons among diffusion variables of 18 models.

Network	PNE		Saturation Percentage		Saturation Time		Win Count	
			Float	Fix	Float	Fix	Float	Fix
Facebook	Leader	M (SD)	.99 (.09)	.83 (.12)	6.21 (1.89)	3.96 (.57)	55.47 (32.51)	55.34 (32.50)
	SE	M (SD)	.99 (.09)	.77 (.11)	6.21 (1.89)	3.96 (.57)	55.47 (32.51)	55.17 (32.31)
	Time	M (SD)	.85 (.11)	.80 (.11)	4.83 (.68)	4.06 (.11)	55.46 (32.53)	55.17 (32.33)
Advice	Leader	M (SD)	.99 (.11)	.83 (.11)	2.26 (.41)	1.95 (.37)	37.83 (22.17)	37.83 (22.17)
	SE	M (SD)	.99 (.11)	.70 (.09)	2.26 (.41)	1.95 (.33)	37.83 (22.17)	37.83 (21.85)
	Time	M (SD)	.97 (.12)	.73 (.12)	4.84 (1.37)	2.86 (.36)	37.83 (21.96)	37.82 (21.98)
Jazz	Leader	M (SD)	.89 (.10)	.78 (.09)	5.40 (.74)	4.36 (.42)	98.46 (56.98)	98.20 (56.76)
	SE	M (SD)	.89 (.10)	.72 (.08)	5.40 (.74)	4.36 (.43)	98.46 (56.98)	98.11 (56.17)
	Time	M (SD)	.84 (.10)	.76 (.09)	5.98 (.76)	4.50 (.41)	98.46 (56.89)	98.25 (56.90)
		Grand Mean	.93 (.10)	.77 (.10)	4.28 (.99)	3.55 (.39)	63.91 (37.20)	63.74 (37.00)

Table 5. Correlations among PNEs.

Saturation Percentage									
	Facebook			Advice			Jazz		
	L	SE	T	L	SE	T	L	SE	T
L	-	.70	.71	-	n.s.	n.s.	-	.85	.91
SE	.93	-	.98	.84	-	.76	.91	-	.96
T	.97	.96	-	.93	.68	-	.95	.97	-
Saturation Time									
	Facebook			Advice			Jazz		
	L	SE	T	L	SE	T	L	SE	T
L	-	.58	.65	-	.29	n.s.	-	.62	.65
SE	1.00	-	.94	.98	-	.51	.99	-	.75
T	.98	.98	-	.62	.68	-	.95	.95	-
Win Count									
	Facebook			Advice			Jazz		
	L	SE	T	L	SE	T	L	SE	T
L	-	.97	.99	-	.97	.98	-	.95	.97
SE	.98	-	.99	.99	-	.98	.97	-	.99
T	.99	.99	-	.99	.99	-	.98	.99	-

Note. Upper diagonal = floating model; lower diagonal = fixed model; n.s. = not significant

References

Abrahamson, E., & Rosenkopf, L. (1997). Social network effects on the extent of innovation diffusion: A computer simulation. *Organization Science*, 8(3), 289–309.

Anderson, P. (1999). Complexity theory and organization science. *Organization Science*, *10*(3), 216–232.

Axelrod, R. (1997). The dissemination of culture: A model with local convergence and global polarization. *The Journal of Conflict Resolution*, *41*(2), 223–226.

Bass, F. M. (2004). New product growth for model consumer durables. *Management Science*, *50*(*12*), 1825–1832 (original work published 1969).

Becker, M. H. (1970). Sociometric location and innovativeness: Reformulation and extension of the diffusion model. *American Sociological Review*, 35, 267–282.

Bonabeau, E. (2002). Agent-based modeling: Methods and techniques for simulating human systems. *Proceedings of National Academy of Science of the United States of America*, 99(3), 7280–7287.

Borgatti, S. P., Everett, M. G., & Freeman, L. C. (2002). UCINET for Windows: Software for social network analysis. Harvard, MA: Analytic Technologies.

Burt, R. S. (1987). Social contagion and innovation: Cohesion versus structural equivalence. *The American Journal of Sociology, 92*(6), 1287–1335.

Carley, K. (1991). A theory of group stability. *American Sociological Review, 56*(3), 331–54.

Coleman, J. S., Katz, E., & Menzel, H. (1957). The diffusion of an innovation among physicians. *Sociometry, 20*, 253–270.

Cross, R., & Parker, A. (2004). *The hidden power of social networks*. Boston, MA: Harvard Business School Press. Data available at http://toreopsahl.com/datasets.

DeBresson, C., & Amesse, F. (1991). Networks of innovators: A review and introduction to the issue. *Research Policy, 20*(5), 363–379

Delre, S. A., Jager, W., & Janssen, M. A. (2007). Diffusion dynamics in small-world networks with heterogeneous consumers. *Computational & Mathematical Organization Theory, 13*(2), 185–202.

Garcia, R. (2005). Uses of agent-based modeling in innovation/new product development research. *The Journal of Product Innovation Management, 22*, 380–398.

Gleiser, P. M., & Danon, L. (2003). Community structure in jazz. *Advances in Complex Systems, 6*, 565–573. Data available at http://deim.urv.cat/~aarenas/data/welcome.htm.

Granovetter, M. (1987). Threshold models of collective behaviors. *The American Journal of Sociology, 83*, 1420–1443.

Guardiola, X., Díaz-Guilera, A., Pérez, C. J., Arenas, A., & Llas, M. (2002). Modeling diffusion of innovations in a social network. *Physical Review E, 66*, 026121–1-4

Katz, E. (1957). The two-step flow of communication: An up-to-date report on a hypothesis. *Public Opinion Quarterly, 21*, 61–78.

Macy, M. W., & Willer, R. (2002). From factors to actors: Computational sociology and agent-based modeling. *Annual Review of Sociology, 28*, 143–166.

Reynolds, C. W. (1987). Flocks, herds, and schools: A distributed behavioral model. *Computer Graphics, 21*, 25–34

Rogers, E. M. (1995). *Diffusion of innovation*. New York: Free Press.

Rogers, E. M., & Beal, G. M. (1958). The importance of personal influence in adoption of technological changes. *Social Forces, 36*(4), 329–334.

Valente, T. W. (1995). *Network models of the diffusion of innovations*. Cresskill, NJ: Hampton Press.

Valente, T. W. (2005). Network models and methods for studying the diffusion of innovations. In P. J. Carrington, J. Scott, & S. Wasserman (Eds.), *Models and methods in social network analysis* (pp. 87–116). New York: Cambridge University Press.

Valente, T. W., Coronge, K., Lakon, C., & Costenbader, E. (2008). How correlated are network centrality measures? *Connections, 28*(1), 16–26.

Valente, T. W., & Davis, R. L. (1999). Accelerating the diffusion of innovations using opinion leaders. *The ANNALS of the American Academy of Political and Social Science, 566*, 55–67.

Wasserman, S., & Faust, K. (1994). *Social network analysis: Methods and applications*. Cambridge, UK: Cambridge University Press.

Watts, D. J., & Dodds, P. S. (2007). Influentials, networks, and public opinion formation. *Journal of Consumer Research, 34*, 441–458.

Weimann, G. (1982). On the importance of marginality: One more step into the two-step flow of communication. *American Sociological Review, 47*, 764–773.

Weimann, G. (1994). *The influentials: People who influence people*. Albany, NY: State University of New York Press.

Wejnert, B. (2002). Integrating models of diffusion of innovation: A conceptual framework. *Annual Review of Sociology*, *28*, 297–326.

Young, P. H. (2006). The diffusion of innovation in social networks. In L. E. Blume & S. N. Durlauf (Eds.), *The economy as an evolving complex system III* (pp. 267–281). New York: Oxford University Press.

· 8 ·

Media Substitution

Supplementation, Complementarity, or Displacement?

Carolyn A. Lin

Introduction

When forecasting the media substitution effects stemming from the competition between old media and new media, some scholars emphasize technology determinism and others focus on social-scientific positivism. Industry experts often paint an optimistic picture of the new media and prophesize impending doom for the old media. As changes in cultural values, social fads, and lifestyle trends often come into play, failures in technology forecasting are as common as the last generation of discarded cell phones.

Three factors—superior content, technical benefits, and cost efficiency delivered by a new technology—in conjunction with the technology's social or lifestyle character typically help determine the new favorites in the technology marketplace. In cases where a drastic gap exists in the evaluation of these four factors between a new and an existing technology, a displacement effect can take place. When a new technology compares favorably with an existing technology in only some of these four factors, a supplementation or complementarity mechanism can occur. An existing technology could sustain itself by expanding its technical capabilities to adapt to audience preferences and lifestyle changes. Alternatively, an existing technology could also remain relevant by reinventing itself with updated technical features and repositioning itself with a new identity.

Background

Throughout the history of mass media, technological innovations and market competition have been the driving forces behind the evolution and substitution in media content and delivery channels. While motion pictures and radio widened audience reach by enabling non-readers to access media offerings, the advent of television accentuated a ubiquitous mass medium—one that is capable of informing and entertaining the masses as well as cultivating and reflecting our popular culture. As the story goes, television became the most popular mass medium. Given television's ability to be the most accessible storyteller with a mass appeal that can transcend national boundaries and divergent cultures, it is not surprising that TV remains the only mass medium that has earned the enviable audience affinity characterized by the "appointment TV" idiom.

When audiences were introduced to VCRs, cable television, and satellite, their control over their viewing schedule and content choice also increased in an unprecedented manner. The term "appointment TV" hence took on an additional meaning, wherein the audience could determine their appointment timing. Moreover, the net effect of this increased content delivery-channel diversity was obvious, as television and other traditional media began to alter the substance of their content and the way they deliver their content and to their audiences. The arrival of the digital or computer-mediated communication age, highlighted by the rise of the Internet, telephony, and mobile communication, brought yet another monumental evolution in the ways that people consume media as well as send, receive, and share information.

The Internet and wireless communication gained popularity within a decade since their introduction to the wider consumer market in the mid-1990s. Since then, the march of the digital age inevitably forged a technology convergence between traditional and online media delivery platforms. Nonetheless, the traditional media have been slow in grappling with the economic potential of such technology convergence, as well as its impact on audience fragmentation and media substitution phenomena. Even as print media have been in steady decline in the past decade, the print media industry has yet to generate a survival strategy that is economically viable for the long-term.

In the new decade of 2010, while mobile media access and on-demand content delivery are the holy grails the media industry is desperately seeking, the Internet, television, radio, and print media market landscapes are also changing rapidly. As the conventional boundaries between traditional and digital media continue to blur, the phenomenon of media substitution in content and delivery channel preference has also become complex. Even though the concept of media substitution is not necessarily new, the reasons and processes behind each media substitution occurrence are often unique and carry strong implications for the life span of the new media and their potential media substitution targets.

Theoretical Perspectives

Several major theoretical perspectives have been utilized to examine the phenomenon of media substitution. They generally fall into the following three categories:

1. An ecological perspective, where different media alternatives within the media environment compete for the same and fixed audience leisure-time resources;
2. A functional perspective, where a media option that is deemed as a desirable functional alternative can replace its functionally similar counterpart;
3. A compensatory perspective, where media-use preference is determined by comparing the content, delivery channel, and cost factors associated with different media options.

Ecological Perspective

The theory of niche has been an important ecological concept for studying how different populations or entities compete for the same resources in the same environment for survival purposes for over a century. Grinell (1917) first introduced the term niche to describe an assortment of environmental conditions under which a species may exist. Hutchinson (1957) expanded the niche concept by proposing that the niche is a hyperdimensional space containing a set of environmental conditions where a species can exist and such space can be reduced when interactions with other species compete for the space. This niche concept was adopted by several disciplinary areas that are relevant to the current discussion, i.e., marketing, organizational, and media research.

For instance, in an organizational setting, an organization as an organism needs to compete effectively in utilizing the same niche environmental resources to survive, and those that are incapable of coping with the sharing or securing the necessary niche resources will wither (Hannan & Freeman, 1989). For example, media content and delivery channels are products that sustain the media organizations, and they often compete for the same niche resources in the same media market.

As the environmental niche changes, a population needs to adapt to the changes in order to survive within the limited niche resources or space. While a population may completely transform itself or divide itself into sub-populations to compete for its possession of the niche, other populations may converge on the same niche to compete for the niche and try to eliminate the competition (Wilson, 1992). Following the logic of the niche theory, when two populations that utilize the same resources to survive are competing to occupy the same resource niche, the likely outcomes could include: (1) partitioning each population's share of the niche resources or (2) having one of the populations be excluded from occupying the same niche resources.

Gratification opportunity theory, as proposed by Dimmick and Rothenbuhler (1984), is a perspective that bases its theoretical rationale on the niche theory and the uses and gratifications perspective (Rubin, 1994). It suggests that audiences make media consumption decisions based on the perceived gratification opportunities/utilities that are available to them (Dimmick, Patterson & Albarran, 1992). When two media choices provide the same gratification opportunity or the same gratifications that they expect to receive, the media-use choices that offer the most desired perceived gratification obtained will either be chosen more often or completely displace their weaker counterparts (Dimmick et al., 2000).

As such, these perceived gratifications obtained (e.g., entertainment, diversion, surveillance, etc.) present themselves as gratification opportunities for the audience and can be considered utilities or niche resources that will be evaluated by the audiences. The audiences' media-use decisions then reflect the competition outcomes between these gratification choices in terms of the location or space that they occupy in the audience leisure-space niche. Some of these gratification opportunities may share the niche space, while others will be excluded.

Functional Perspective

Media-use stimulated gratifications is a multidimensional concept, as it can reflect a range of cognitive, emotional, or behavioral responses to media content that stem from a set of psychological antecedents. For instance, media coverage of the devastation caused by the 2010 earthquake in Haiti has helped unveil the island nation's impoverished economic status, elicit sympathy toward the victims, and draw financial resource support at a global level. In other words, audiences from around the world have been informed and touched as well as taken action in response to this unfortunate human condition. While television programs such as *Desperate Housewives* provide an escape for the audience, online social networks such as Facebook are favored by the users because they can provide a safe and emotionally intimate venue for seeking/establishing social identify and companionship. These media use examples provided a number of utilitarian functions that met a set of audience psychological needs.

However, when a new or alternative media option could provide the same function to meet these psychological needs, the audience may choose to adopt this new option over its older counterpart. This phenomenon is characterized as the displacement of an old media choice by a functionally equivalent new media choice. The principle of functional equivalence was introduced by Himmelweit, Oppenheim, and Vince (1958), who tried to explain how the television medium's popularity had adversely affected audience time devoted to the use of other media. For instance, the radio used to be the primary source for the audience to seek new information, companionship, and escape—as well as excitement and entertainment—before it was displaced by television (e.g., Coffin, 1955; Parker, 1961).

In retrospect, the most popular radio program genres in the 1950s were transplanted onto the television screen. These included, for example, soap opera serials, variety shows (that contained live music and comedic performers, etc.), and dramas. The audiences quickly realized that television was a better medium for delivering these types of programs, owing to its ability to present the content of these programs with compelling visuals. Considering how humans are capable of processing visual information more readily and effectively than audio-only information in a mediated communication setting, television's ability to allow us to share the human experiences on screen gave "seeing is believing" a new meaning. Needless to say, the role of television in influencing audiences' perceived social reality—as well as an array of other socio-cultural beliefs, attitudes and behaviors—remains the corner stone of mass communication research until today.

This functional-equivalence displacement mechanism has also been examined in the dimension of time and financial resources via the principle of relative constancy. The principle of relative constancy, as defined by McCombs (1972, p. 10), states that "The level of spending on mass media by consumers and advertisers is determined by the general state of the economy. Any change in the level of the economy causes a parallel change in spending in mass media." This principle has been applied to examine the ways audiences allocated their money and time in their media-leisure pursuit, under the assumption that any changes in media choices and consumption are reflective of a reshuffling of socio-economic resources that will result in resorting the audience media-use resources allocation to a fixed constant (McCombs & Nolan, 1992; McCombs, 1993).

As the media environment changed with the introduction of VCRs and cable TV, the empirical validity of the principle of relative constancy has been called into question by some (Grant, 2007). Analyses of consumer media expenditures show that even though mass media spending is largely discretionary and it can be influenced by the general economic conditions in society, there is no one-on-one relation between these two economic indicators (Dupagne, 1997). For instance, national GDP was found to be positively and significantly correlated with the use and adoption of television, mobile phones and the Internet (Nam & Barnett, forthcoming). Lacy and Noh (1997) further pointed out that the ambiguity of what is supposed to be held constant and the lack of consideration of the price of the good has muddled the empirical validity of the relative constancy assumption—in predicting the media displacement phenomenon—as related to functional equivalence. In point of fact, when new media technologies were introduced, consumers did reallocate some of their media choices, but they also increased their media spending to accommodate the adoption of new media options (Hindman & Wiegand, 2008). For example, movie attendance was adversely affected by the introduction of the VCR, but the popularity of the VCR and cable TV did result in increased media spending (Lacy & Noh, 1997).

Compensatory Perspective

The two theoretical approaches discussed above each concentrate on a particular aspect of the media use phenomenon. In particular, the ecological perspective, which has been examined primarily with applying the gratification opportunity theory, focuses on the media gratification niches and how they displace or share the space in the audience's media-use repertoire. By contrast, the functional approach emphasizes the mechanism where the new or alternative media options can displace their older or existing functionally equivalent counterparts, whether they be content or delivery channel preferences. Neither perspective specifically addresses the media displacement phenomenon, which involves a combination of content preferences, perceived channel attributes, technology adoption factors, and material resource considerations (e.g., time and money).

To address the multidimensional nature of the media substitution phenomenon, an integrated approach is presented here to incorporate the main theses of the existing conceptualizations and additional theoretical elements to provide an alternative analytical framework. This integrated approach reflects a *relative compensatory perspective*, which asserts that when a new content choice and/or delivery channel is preferred over the existing content and/or delivery channel selection, several conditions will need to be met before any degree of displacement between the existing and new options will take place. These conditions can include content superiority, technical superiority, and greater cost efficiency (Lin, 2001a). Depending on which one or combination of these three conditions are met, the media selection outcomes will differ and can result in media supplementation, complementarity, substitution, or displacement. The mechanism of this media selection process is characterized by how the audience compensates the lack of access to a certain media resource by securing an attainable alternative to optimize their media-use goals. For example, if an audience member deems that watching movies via pay-per-view cable services is not economically viable, he may compensate for the lack of ready access by becoming a member of Netflix (to receive their preferred viewing choices through this DVD rental mail service) to achieve the same media consumption objective.

To further elucidate this particular perspective, it is important to first conceptually decipher, define, and differentiate the different terminologies that have been loosely used in the research literature on this topic. The term "media displacement" has been widely used to describe the phenomenon where the audience displaces a media content choice or content delivery channel with another one. If we examine the displacement concept more closely, there seems to be an implied semantic and connotative absolutism whenever an act of "displacement" is being enacted. The definition of "displacement" often refers to a physical replacement of substance (e.g., atoms), direction (e.g., energy movement), or space (e.g., fluid or geological fault) as well as a transfer of emotions or behavioral response. As pure displacement of media content rarely took place, we should consider conceptual implications associated with the use of this term. This is evidenced by the fact that most means of communication, whether they be mediated or non-mediated, have stood the test of time in one form or another.

For instance, smoke signals are still being used for sounding alerts when all else fails. Digital video discs (DVDs) are the new faces for their bulky optical disc precursors. The fading VCR, however, is an example of hardware displacement, where a piece of analog hardware is being displaced by other functionally equivalent digital hardware (e.g., DVDs) to provide the same content delivery services. As for media content displacement, it is safe to say that all forms of television program genres developed since the inception of television broadcasting are still in place today. The same is true for the radio program genres, where some of them disappeared from the radio (e.g., soap operas) but have survived on television.

A more inclusive way to capture the uneven gradients of this phenomenon, a more broadly scoped term—media substitution—would seem to be more closely reflective of its complex conceptual dynamics. Semantically speaking, the definition of substitution implies a less permanent or absolute form of replacement; a substitution often takes the form of short-term replacement or time-shift deployment between two components of a group (e.g., temporary replacement or turn-taking between team members) or a fluid replacement of functionally equivalent elements (e.g., mathematical elements or logical expressions). In other words, the media substitution concept seems to mirror the phenomenon where temporary replacement or transformational replacement fluidity in the use of different media content or delivery channels takes place.

Based on this conceptual distinction, displacement refers to a permanent replacement of media resources, and substitution would suggest a more flexible replacement of these media resources. As absolute media displacement is not common, as such, there are also two additional variations of the substitution concept: media supplementation and media complementarity. These two ancillary concepts are explicated below.

The conceptual framework of media supplementation illustrates the scenarios where one or more media content platforms and/or delivery channels are used to provide access to additional media content platforms/delivery channels, resulting in supplementing the use of another primary media content platform/delivery channel. The ultimate function of this supplementation mechanism is to make both the media content and channel use experience more complete or exhaustive. For instance, avid news consumers are usually regular newspaper readers. Yet they are often the loyal readers of news magazines as well. News magazine use, then, can be considered as providing supplemental content to newspaper reading by adding value to the news consumption behavior and making the news consumption experience more complete. Another example could involve accessing additional news coverage through a completely different delivery channel such as the Internet. Accessing online news reports posted by various offline news media sources (e.g., CNN, *The New York Times*) and content gatherers (e.g., Yahoo News, The Huffington Report) is another means of supplementing one's news consumption.

By contrast, the function of media complementarity is usually materialized by utilizing one or more different types of media content and/or delivery channels to complement the use of another primary media content/delivery chancel choice, resulting in improved efficiency of one's media-use process without gaining access to additional media content platform/delivery channels. For instance, television viewers often program their digital video recorders (DVRs) to cast their preferences in terms of program choice and viewing schedule. In doing so, they are using the DVR to complement the use of the television medium by increasing their control over their viewing process and enabling more effective utilization of the existing media content/delivery channel resources. Another example could involve how Internet users typically download music via a personal computer and then onto a hand-held mobile device, so that they could access music files on the go. In this way, they complement the use of a personal computer with a limited-function hand-held mobile computer to enhance their control over their physical access to the music content and the timing of such access.

Empirical Research

A large number of studies have examined the media substitution phenomenon with various empirical approaches and methodologies. While the early research on this topic tended to be descriptive in nature, more recent studies have provided research findings that are more theory-driven and explanatory in their endeavors. The review presented below will illustrate the three evolutionary landmarks of the media substitution phenomenon to provide the historic socioeconomic contexts in which the new media content and delivery modalities were introduced to the consumer. These three evolutionary landmarks—the television age, the video age, and the Internet age—are not necessarily mutually exclusive, as they each reflect a point in time during which the development of their respective technological modality has first secured their place in media history.

The Television Age

Television provided us with the first widely noted media substitution phenomenon on the basis of functional equivalence. Radio was substituted by television as the most dominant entertainment medium for American families (Lazarsfeld & Kendall, 1948), as family members gathered around the television set to watch an assortment of entertainment programs including music/variety shows, serial dramas, soap operas, sports coverage, and news instead of tuning into these programs on the radio. Television provided the audience an unprecedented opportunity to enjoy audiovisual entertainment and news reports in the privacy of their own home—an activity that used to be the exclusive domain of the movie theaters.

Radio listening time took a severe dive, when Middle America embraced their television sets as the core mass medium for family entertainment (Parker, 1961; Schramm, Lyle & Parker, 1961). Over time, radio-listening level made a gradual comeback, when the leisure time allocated between television and other media became more stabilized (Bogart, 1956). Young children's radio listening time, however, never did return to its previous level and remains low, as few radio programs target their age group (Himmelweit et al., 1958). Other casualties of the television medium's popularity included reduced amounts of audience time spent with movies, books, magazines, and newspapers (Brown, Cramond & Wilde, 1974; Weiss, 1969). Studies have also found that young children had spent less time outdoors playing, owing to their television viewing habit (Maccoby, 1951).

Radio's place in the audience media-leisure space eventually settled in the position as the primary niche source for music entertainment as well as talk programs (since the late 1980s, when the fairness doctrine still stood as an FCC regulation). Similarly, even though movie attendance declined drastically for a period of time (Belson, 1967), it regained its footing with the adolescent population after it established its television viewing pattern (Comstock, Chaffee, Katzman, McCombs & Roberts, 1978). The long-term impact of television on adult reading habits was mixed (Brown et al., 1974; Coffin, 1955; Schramm et al., 1961). In recent years, the publishing industry is still flush with best-selling popular titles. The market shares for printed magazines and newspapers are less certain, as cable television and Internet media have been successful in competing for advertising revenue from these print media outlets.

It seems that the television medium created two types of media substitution outcomes. One, television had an impact on a functional-equivalence substitution effect on radio in that there was very limited overlap in the entertainment programs aired between these two media delivery channels, after most entertainment content genres on radio were transplanted on to television. Television content, however, may have a time displacement effect on the patronage of newspapers, magazines, books or movies. This suggests that increased time spent on television viewing reduced the amount of time devoted to these other mass media, albeit in a non-linear fashion. Children read less than before, but adolescents did not attend fewer movies. The newspaper, magazine, and book publishing industries thrived instead of declining, until electronic and digital media delivery channels started to compete for audience leisure time.

More important, the key socio-psychological reason that has enabled television to become the primary functional alternative to all other traditional mass media is the perceived psychological gratifications that its vast audiences can expect to receive (Dimmick, Patterson & Albarran, 1992). These perceived media gratifications often reflect how the television medium could satisfy such audience needs as news surveillance, information seeking, excitement seeking, escape, diversion, relaxation, companionship, passing time, social identity, and para-social interaction (Rubin, 1994). With such wide array of psychological need that the television medium could potentially fulfill for its viewing audience, television's status as the dominant leisure-time media choice has long been established until another media venue can successfully challenge it as a functional-alternative must.

From a technological point of view, television's staying power as the reigning mass entertainment medium has also been facilitated by the continuing innovations that have improved its ability to provide content superiority, technical superiority, and greater cost efficiency (Lin, 2001). For instance, cable television systems began to provide channels numbered in the dozens in the late 1980s. Since the late 1990s, digital cable television and digital satellite television services have offered television viewers with over 100 channels of programming. This channel diversity has greatly enhanced the ability of the television medium to transit itself into the Internet age without falling behind the innovation curve and the audiences' desire to access niche program content on demand (Lin, 2004).

The Video Age

The video age arrived in the mid-1970s, when Betamax, VHS and LaserDiscs (LDs) were introduced in 1975, 1976, and 1978, respectively. Betamax, with its 60-minute recording time, had a higher recording quality than that of VHS, which could record up to three hours; the LDs were a superior but expensive playback-only device. As more movie studios favored the VHS's ability to store a full-length feature film and its lower consumer cost than LDs, the consumers followed suit. The net outcome of this content, technical delivery, and cost competition is a compensatory displacement dynamic, where the consumers chose to compensate the loss of higher recording quality and replay quality (at a higher cost) with VHS's longer recording time and lower cost. Even though the analogue LDs never gained popularity, they were the precursors of the digital Compact Discs (CDs) and Digital Video Discs (DVDs).

The videocassette recorder (VCR) had a transformational effect on how the audience allocated its leisure time, in much the same way that television did over two decades earlier (Atkin, Greenberg & Baldwin, 1991; Klopfenstein, 1989; Levy, 1987, 1989). In particular, as a majority of American homes owned more than one TV set, the VCR created a home video culture by bringing back family viewing and social viewing to a home leisure setting (Gunther & Wober, 1989; Holman, 2000; Levy & Gunther, 1988; Lin, 1992, 1994a; Lindlof & Shatzer, 1990). Family and friends engaged in interpersonal communication about what videos to rent and when to watch them together (Rubin & Eyal, 2002). VCR use was also found to increase audiences' TV viewing gratifications (Lin, 1990; Rubin & Rubin, 1990) and as a functional alternative to interpersonal communication (Rubin & Bantz, 1987).

While the VCR is being used to record television programs for time-shifted viewing, building a video library for future viewing and playing back video rentals, the VCR does not displace television viewing. Instead, it functionally complements the television medium to enhance audience control over their television viewing schedule. Even as the VCR has had a functional substitution impact on the movie-going activity (Henke & Donohue, 1989; Lin, 1992; Noh & Grant, 1997), it does not displace the movie-viewing activity and hence serves a functional complement to allow the audience to make up for their lack of time or financial resources to attend movie theater screenings (Krugman & Johnson, 1991; Lin, 2001a).

The Sony Walkman, hailed as one of the most significant inventions of its time (Patton, 1999), was undoubtedly the first personal analogue media product that allowed its users to have sole control over its applications. It was displaced by the compact disc (CD) players, which compensated for the Walkman's limited content capacity, technical quality, and cost efficiency with greater content capacity, superior sound quality, and slightly higher price. The CD's compensatory functional displacement not only retired the analogue/digital audiocassette recorders and the Walkman to a technology museum, but also helped launch the first widely embraced personal digital media product in history.

While CDs don't have nearly enough memory capacity to store feature-length films, digital video disc (DVD) players emerged in 1997 to do just that in order to compete with VCRs. Since the early 2000s, DVD players have largely displaced the analogue VCRs for video-replay functions. Similarly, digital video recorders (DVRs), first appearing in 1998, also have successfully hampered the market share of the DVD recorders during the mid-2000s, owing to its efficient and superior recording and replaying mechanisms. In both of these cases, content capacity and technical programming superiority helped trigger such a functional displacement outcome between the winners and losers in the home entertainment technology niche space.

Early evidence of DVD's technical displacement function over VCRs was obvious; what was less clear was DVD's intrusion on other media-use activities (Kim & Lee, 2003). This result was not surprising, as DVDs were used primarily to displace the VCR's replay functions; they did not physically alter the amount of time allocated for other media use. By the same token, DVR's functional displacement of VCRs also did not signify a time displacement. Hence, the three major players of the video age—VCRs, DVDs, and DVRs—can all be considered as functionally complementary to the television medium itself. A complete technical displacement of DVRs over VCRs will occur when the VCR's utility—in providing pre-recorded content and video replays—is no longer valuable to the audience in the Internet age. In the Internet age, where technology fluidity is the key that allows the audience to access media content anywhere anytime (Lin, 2004), all analogue media technologies will become collector's items in good time.

The Internet Age

Several notable prototypes paved the way for the Internet to become universally available to all able users. For instance, consumer videotext services

(Ettma, 1984; Lin, 1994b) such as Viewtron (1983–1986), which provided news, local sports, banking, police reports, weather, travel booking, and shopping services, etc., was the content template for the commercially oriented online services today. CompuServe, the first wide-area online service adopted by corporate and institutional users (in 1982), was later purchased by AOL (in 1998), which utilized CompuServe's dial-up network infrastructure to launch the first successful consumer-oriented online service. The ListServ, widely adopted since 1986, offers new-generation service that's still in use today by Internet users from around the world. While the technologies of these rudimentary online services have been integrated into their modern-day equivalents, this particular media substitution phenomenon marks another example of progressive technical substitution instead of complete displacement.

The Internet technology in the hypertext markup language (HTML) environment has lived up to its name as the World Wide Web, with countless individuals, groups, organizations, and institutions (including commercial media) constantly generating and sharing a vast amount of diverse content from all corners of the world. Equally omnipresent is the number of people who access the Internet on a daily basis. Today, the Internet has become the biggest single repository of information and interpersonal communication channels (e.g., email systems and Facebook); its entertainment content has also been expanding rapidly (e.g., Hulu), especially through user-generated content (e.g., YouTube). As such, the Internet has also been considered a serious competitor or media substitute for traditional media, as they rely on the same audience constituencies, subscription fees and/or advertising revenues to sustain their economic base.

Existing empirical research that examined this emerging media substitution dynamic between the use of online media content and other media content has generated mixed results. For instance, some studies found the Internet to be a complementary medium to traditional media when it comes to audience access time and consumption purposes (Coffey & Stipp, 1997; Robinson & Kestnbaum, 1999). Other researchers report that audiences reallocated their leisure-time resources by expending them on the Internet at the expense of traditional media (James, Worting & Forrest, 1995; Kaye, 1998). These inconsistent findings could probably be partially explained by the differences in research methodologies, sample compositions and measurement used as well as the theoretical frameworks from which the research was based. To wit, college students' Internet and media-use patterns are not necessarily representative of those of the general public. Qualitative and quantitative study results should not be interpreted with the same analytical criteria; the same is true for descriptive and explanatory research outcomes.

As a case in point, some studies measured time displacement between Internet use and other media through descriptive measures, including a simple correlation, without exploring the contextual reasons why the audience spent more time on one medium versus another. These studies often did not take into account the fact that the Internet is a tool used for work and play, unlike other traditional media outlets. Moreover, depending on the time period during which the study was conducted, the Internet could either be a novelty or a necessity in terms of the niche it occupies in the audience's time spent with media on the whole. It is also useful to acknowledge that the Internet remains a young and evolving medium and its relations with the audiences are still emerging. As the Internet is a medium whose content and uses are shaped by the established content providers as well as the audiences themselves, the Internet use phenomenon could also shift and be reshaped quickly.

An early study of online bulletin use (James, Worting & Forrest, 1995), for instance, found a time displacement effect in accessing the electronic bulletin board on other media use; the non-representative sample (9% response rate) should be taken into account when evaluating the study results. Another study (Robinson, Barth & Kohut, 1997) compared audience time allocation across computer-mediated communication and other media (between 1994 and 1995) reported a supplemental relationship instead of time displacement between them. Both studies were conducted with early Internet adopters, who tended to be researchers, institutional and corporate users, before AOL made online services widely accessible to the general public (since 1996).

Research findings from the next phase of the Internet age, where online service adoption was beginning to reach a critical mass, were equally mixed. For instance, Kayany and Yelsma's (2000, p. 223) investigation of family units showed that the time displacement levels ranged between "decreased somewhat" and "remained the same," even though a significant time displacement effect on telephone use, newspaper reading, and family conversations was found between heavy and light online users. While Nie's (2001) national survey sample showed that increased time with Internet use meant decreased time for social interactions his claim was rejected by Neustadtl and Robinson (2002), who dismissed the notion that undifferentiated Internet use led to lower social involvement, based on the General Social Survey results. Althaus and Tewksbury's (2000) college student sample indicates that while there was no displacement effect between accessing the Internet as a news source and television news viewing, seeking news online was related to higher levels of offline newspaper reading. Likewise, Lin's (2001) general adult population study suggests that online media access activity appeared to be supplemental to offline media consumption.

The research evidence during the past few years revealed additional trends not addressed during the decades prior, as web use has become a daily routine for most users and personal web applications also began to multiply. For instance, Cai (2004) asked college students to give up computer use (except for completing homework assignments) for one day and television viewing for another day. Students did not increase time spent with traditional media when computer access was banned, but they spent more time with newspapers and radio (instead of the computer or other media) when television viewing was abolished. According to these findings, there was no functional equivalence mechanism at play to generate either a time or functional equivalence displacement effect. By contrast, Valkenburg and Peter (2007) investigated 1,210 Dutch young people's (aged 10–17) use of the Instant Messaging (IM) service in relation to stimulation versus displacement of social relations. Their study outcomes supported that IM use stimulated rather than displaced time spent with their existing friends and the quality of such friendships.

Tsao and Sibley (2004) investigated the potential functional equivalence effect between Internet advertising and traditional media advertising with a general adult population sample. They found that while Internet and traditional media advertising shared complementary and reinforcement (increase-increase) effects, this symbiotic relation did not occur with free community newspapers and weekly newspapers—whose advertising niche showed a functional displacement tendency relative to Internet advertising. Okazaki and Hirose (2009) validate the theoretical frameworks of gratification opportunities (Dimmick et al., 2004), complementary relations between online and traditional media preference (Lin, 1999), and reinforcement effects between online and traditional media advertising (Taso & Sibley, 2004) in their national study, which examined the functional equivalence relations between and consumer involvement with PC-based Internet, mobile Internet, and traditional media. Their findings suggest a complementary effect between consumer attitudes toward traditional media, PC-based Internet, and mobile Internet, consumer satisfaction with one medium could influence the gratification opportunities for other media, and mobile PC users will reinforce their information search via PC-based Internet for a high-involvement product.

Implications for Future Research

The theoretical perspectives and the empirical evidence discussed in the current chapter have provided a useful framework to help explore the ever-changing relations between content invention/reinvention, technological innovations and the media substitution phenomenon. It would seem that the relations between the digital media and their audiences are fertile ground for researchers, academic or otherwise. For instance, IM has been mostly displaced in the past few years by social MySpace, when the latter became all the rage. In 2010, Facebook and Twitter have made MySpace the less attractive social networking choice. What are the effects of the popularity of online social networks and user-generated content on other media content or technology use?

In order to predict what the future media substitution dynamics might realize, it is useful to develop a set of factors that could help pinpoint such a forecast below. Generally speaking, the acceptance and popularity of contemporary entertainment and communication media technologies are often determined by several major factors. These factors can include: (1) whether the functions, applications, and content provided by the technology can accommodate a user-centered design in terms of their usability and usefulness, (2) whether the user interface with the technology and the content that it delivers is deemed useful to meet the audience needs for completing communication task and leisure pursuits, and (3) whether the audiences' psychological needs can be sufficiently satisfied to encourage audience adoption, sustain audience patronage, and secure audience loyalty.

In sum, three objective criteria—content superiority, technical superiority, and cost efficiency delivered by a new technology—typically help determine the new favorites in the media technology marketplace. This occurs in conjunction with two subjective criteria: the technology's social attribute (e.g., perceived as a social medium) or lifestyle character (e.g., perceived as an on-demand medium). In cases where a drastic gap exists in the evaluation of these four factors between a new and an existing technology, a displacement effect can take place. In other cases, where a new technology compares favorably with an existing technology in only some of these four factors, a supplementation or complementarity mechanism can occur. Alternatively, an existing technology could sustain itself by expanding its technical capabilities to adapt to audience preferences and lifestyle changes. Similarly, an existing technology could also remain relevant by reinventing itself with updated technical features and repositioning itself with a new identity.

With mobile technologies continuously fusing work, play, and communication functions in a fluid on-demand modality, it would not be overly speculative to predict that the next landmark in the media history will be the mobile media age. If so, what would be the competitive status of the conventional media, video technologies, and PC-based Internet? Even though technology forecasting is a dangerous business, I would venture to say that supplemental and complementary relations between mobile media and other existing media will be the most likely scenario that marks the media substitution outcome, at least in the early phase of the showdown. What I am unable to predict is which medium or media will supplement or complement which other counterparts, mobile or otherwise. Later research should investigate how emerging technology convergence influences the relations between traditional media, digital media, Internet-based media, and mobile media.

References

Althaus, S. L. & Tewksbury, D. (2000). Patterns of Internet and traditional news media use in a networked community. *Political Communication, 17*, 21–45.

Atkin, D., Greenberg, B. S., & Baldwin, T. F. (1991). The home ecology of children's TV viewing: Parental mediation and the new media environment. *Journal of Communication, 41*(3), 88–99.

Belson, W. A. (1967). *The impact of television*. Hamden, CT: Archon.

Bogart, L. (1956). *The age of television: A study of viewing habits and the impact of television on American life*. New York: Frederick Ungar.

Brown, J. R., Cramond, J. K., & Wilde, R. J. (1974). Displacement effects of television and the child's functional orientation to media. In J. G. Blumler & E. Katz. (Eds.), *The uses of mass communications: Current perspectives on gratifications research* (pp. 93–112). Beverly Hills, CA: Sage.

Cai, X. (2004). Is the computer a functional alternative to traditional media? *Communication Research Reports, 21* (1), 26–38.

Carrol, G. R. (1982). *Publish and perish: The organizational ecology of newspaper industries*. Greenwich, CT: JAI.

Coffey, S. & Stipp, H. (1997). The interactions between computer and television usage. *Journal of Advertising Research, 37(2)*, 61–68.

Coffin, T. E. (1955). Television's impact on society. *American Psychologist, 10*, 630–641.

Comstock, G., Chaffee, S., Katzman, N., McCombs, M., & Roberts, D. (1978). *Television and human behavior*. New York: Columbia University Press.

Dimmick, J., Kline, S. & Stafford, L. (2000). The gratification niches of personal e-mail and the telephone: Competition, displacement, and complementarity. *Communication Research, 27*(2), 227–250.

Dimmick, J., Patterson, S., & Albarran, A. (1992). Competition between the cable and broadcast industries: A niche analysis. *Journal of Media Economics*, *5*(1), 13–29.

Dimmick, J. & Rothenbuhler, E. W. (1984). The theory of the niche: Quantifying competition among media industries. *Journal of Communication*, *34*(1), 103–119.

Dupagne, M. (1997). Beyond the principle of relative constancy: Determinants of consumer mass media expenditure in Belgium. *The Journal of Media Economics*, *10*(2), 3–19.

Dutta-Bergman, M. J. (2004). Interpersonal communication after 9/11 via telephone and Internet: A theory of channel complementarity. *New Media & Society*, 6(5).

Ettma, J. (1984). Three Phases in the Creation of Information Inequalities: An empirical assessment of a prototype videotext system. *Journal of Broadcasting* 28(4), 383–95.

Grant, A. (2007). Video technology and home entertainment. In C. Lin and D. Atkin (Eds.), *Communication technology and social change: Theory and implications*. Mahwah, NJ: Lawrence Erlbaum Associates (pp. 167–183).

Grinnell, J. (1917). The niche-relationships of the California thrasher. *Auk 34*, 427–433.

Gunther, B., & Wober, M. (1989). The uses and impact of home video in Great Britain. In M. R. Levy (Ed.), *The VCR age: Home video and mass communication* (pp. 50–69). Newbury Park, CA: Sage.

Hannan, M. T., & Freeman, J. (1989). *Organizational ecology*. Cambridge, MA: Harvard University Press.

Henke, L. L., & Donohue, T. R. (1989). Functional displacement of traditional TV viewing by VCR owners. *Journal of Advertising Research*, *29*(2), 18–23.

Himmelweit, H. A., Oppenheim, A. N., & Vince, P. (1958). *Television and the child: An empirical study of the effect of television on the young*. London: Oxford University Press.

Hindman, D. B. & Wiegand, K. (2008). The big three's prime-time decline: A technological and social context. *Journal of Broadcasting & Electronic Media*, *52*(1), 119–135.

Holman, T. (2000). Foreword. In R. Harley (Ed.), *Home theater for everyone: A practical guide to today's home entertainment systems*. Albuquerque, NM: Acapella.

Hutchinson, G. E. (1957). Concluding remarks. *Cold Springs Harbor Symposium Quantitative Biology*, *22*, 415–427.

James, M. L., Wotring, C. E., & Forrest, E. J. (1995). An exploratory study of the perceived benefits of electronic bulletin board use and their impact on other communication activities. *Journal of Broadcasting and Electronic Media*, 39, 30

Kayany, J. M., & Yelsma, P. (2000). Displacement effects of online media in the socio-technical contexts of households. *Journal of Broadcasting and Electronic Media*, *44(2)*, 215–229.

Kaye, B. K., (1998). Uses and gratifications of the World Wide Web: From couch potato to web potato. *The New Jersey Journal of Communication*, 6, 21–40.

Kim, H. S. & Lee, S. T. (2003). Exploring the Characteristics of DVD home theater system adopters. *Communication & Society*, 6(3), 267–290.

Klopfenstein, B. C. (1989). The diffusion of the VCR in the United States. In M. R. Levy (Ed.), *The VCR age: Home video and mass communication* (pp. 21–39). Newbury Park, CA: Sage.

Krugman, D. M., & Johnson, K. F. (1991). Differences in the consumption of traditional broadcast and VCR movie rentals. *Journal of Broadcasting*, 35, 213–232.

Lacy, S. & Noh, G. Y. (1997). Theory, economics, measurement, and the principle of relative constancy. *The Journal of Media Economics*, *10*(2), 3–16.

Lazarsfeld, P. E., & Kendall, P. L. (1948). *Radio listening in America: The people look at radio*. New York: Prentice Hall.

Levy, M. (1987). VCR use and the concept of audience activity. *Communication Quarterly, 35*, 267–275.

Levy, M. (Ed.). (1989). *The VCR age: Home video and mass communication*. Newbury Park, CA: Sage.

Levy, M., & Gunther, B. (1988). *Home video and the changing nature of the TV audience*. London: IBA & John Libbey.

Lin, C. A. (1990). Audience activity and VCR use. In J. Dobrow (Ed.), *Social and cultural aspects of VCR use* (pp. 93–106). Hillsdale, NJ: Lawrence Erlbaum.

Lin, C. A. (1992). The functions of the VCR in the home leisure environment. *Journal of Broadcasting & Electronic Media, 36*, 345–351.

Lin, C. A. (1993). Exploring the role of VCR use in the emerging home entertainment culture. *Journalism Quarterly, 70*, 833–842.

Lin, C. A. (1994a). Leisure-time reallocation in a video culture. *Mass Communication Review, 21*, 4–16.

Lin, C. A. (1994b). Exploring potential factors for home videotext adoption. *Advances in Telematics, 2*, 111–121.

Lin, C.A. (1999). Online-service adoption likelihood. *Journal of Advertising Research*, 39(2), 79–89.

Lin, C. A. (2001a). The VCR, home video culture and new video technologies. In J. Bryant & J. A. Bryant (Eds.), *TV and the American age* (2nd ed., pp. 91–107). Mahwah, NJ: Lawrence Erlbaum.

Lin, C. A. (2001b). Audience attributes, media supplementation and likely online service adoption. *Mass Communication and Society*, 4(1), 19–38.

Lin, C. A. (2004). Webcasting adoption: Technology fluidity, user innovativeness, and media substitution. *Journal of Broadcasting & Electronic Media*, 48(3), 446–465.

Lindlof, T. R., & Shatzer, M. J. (1990). VCR usage in the American family. In J. Bryant (Ed.), *Television and the American family* (pp. 89–109). Hillsdale, NJ: Lawrence Erlbaum.

Maccoby, E. (1951). Television: Its impact on school children. *Public Opinion Quarterly, 15*, 421–444.

McCombs, M. (1972). Mass media in the marketplace. *Journalism Monographs, 24*, 1–104.

McCombs, M., & Nolan, J. (1992). The relative constancy approach to consumer spending for media. *Journal of Media Economics*, 5(2), 43–52.

Nam, Y. & Barnett, G. A. (2010, forthcoming). Communication media diffusion and substitutions: Longitudinal trends from 1980 to 2005 in Korea. *New Media & Society*.

Nie, N. H. (2001). Sociability, interpersonal relations, and the Internet: Reconciling conflicting findings. *American Behavioral Scientist*, 45(3), 420–435.

Neustadtl, A., & Robinson, J. P. (2002). Social contact differences between Internet users and nonusers in the General Social Survey. *IT & Society*, *1*(1), 73–102.

Okazaki, S. & Hirose, M. (2009). Effects of displacement-reinforcement between traditional media, PC Internet and mobile Internet: A quasi-experiment in Japan. *International Journal of Advertising*, 28(1), 77–104.

Parker, E. B. (1961). Television and the process of cultural change. *Journalism Quarterly*, 38, 39–48.

Patton, H. (1999, July 29). Humming off key for two decades. Retrieved December 12, 2009, from http://www.nytimes.com/1999/07/29/technology/humming-off-key-for-two-decades.html?pagewanted=1.

Robinson, J. P., Barth, K., & Kohut, A. (1997). Social impact research: Personal computers, mass media and use of time. *Social Science Computer Review, 15*(1), 65–82.

Robinson, J. P., & Kestnbaum, M. (1999). The personal computer, culture, and other uses of free time. *Social Science Computer Review, 17*, 209–216.

Rogers, E. (2002). The information society in the new millennium. In C. Lin and D. Atkin, *Communication technology and society* (pp. 22–42). Cresskill, NJ: Hampton.

Rubin, A. M. (1994). Media uses and effects: Uses and gratifications perspective. In J. Bryant & D. Zillman (Eds.), *Media effects: Advances in theory and research* (pp. 417–436). Hillsdale, NJ: Lawrence Erlbaum.

Rubin, A. M., & Bantz, C. (1987). Utility of videocassette recorders. *American Behavioral Scientist, 30*, 471–485.

Rubin, A. M. & Eyal, K. (2002). The videocassette recorder in the home media environment. In D. Atkin and C. Lin (Eds.). *Communication technology and society: Audience adoption and uses* (pp. 329–349). Cresskill, NJ: Hampton.

Schramm, W., Lyle, J., & Parker, E. B. (1961). *Television in the lives of our children*. Stanford, CA: Stanford University Press.

Son, J., & McCombs, M. (1993). A look at the constancy principle under changing market conditions. *Journal of Media Economics*, 6(2), 22–37.

Tsao, J.C. & Sibley, S.D. (2004). Displacement and reinforcement effects of the Internet and other media as sources of advertising information. *Journal of Advertising Research, 44*(1), 126–142.

Valkenburg, P. M., & Peter, J. (2007). Online communication and adolescent well-being: Testing the stimulation versus the displacement hypothesis. *Journal of Computer-Mediated Communication, 12*(4), article 2. Retrieved from http://jcmc.indiana.edu/vol12/issue4/valkenburg.html.

Weiss, W. (1969). Effects of the mass media of communication. In G. Lindzey and E. Aronson (Eds.), *The handbook of social psychology, vol. 5* (pp. 77–196). Reading, MA: Addison-Wesley

Wilson, E. O. (1992). *The diversity of life*. New York: W. W. Norton.

· 9 ·

Turning Diffusion of Innovations Paradigm on Its Head

ARVIND SINGHAL

Introduction

The classical diffusion of innovations paradigm has faced criticism for reifying outside-in, expert-driven approaches to solving problems and for overlooking and rejecting local solutions. In this chapter, we argue that diffusion scholars should pay more attention to approaches such as Positive Deviance (PD) that enable communities to discover the wisdom they already have and then to act on it. PD is an assets-based approach that identifies what's going right in a community in order to amplify it, as opposed to focusing on what's going wrong in a community and fixing it with outside expertise. In the PD approach, the change is led by internal change agents who, with access to no special resources, present the social behavioral proof to their peers that problems can be solved. Given that the solutions are generated locally, and distilled through concrete action steps, they are more likely to be owned by potential adopters and be sustained.

In the past seven decades, since the publication of the Ryan and Gross (1943) diffusion of hybrid seed corn study in Iowa, the classic diffusion of innovations paradigm, and its accompanying practice, is fundamentally premised on the following tenets (Coleman, Katz & Menzel, 1966; Dearing & Meyer, 2006; Rogers, 2003, 2004; Singhal & Dearing, 2006; Valente, 1995):

1. That innovations (new ideas, products, and services) come from the outside,
2. pushed and promoted by a change agency
3. through expert and knowledgeable change agents,
4. who use persuasive communication strategies to plug existing knowledge-attitude-practice (KAP) gaps among the client audience
5. by harnessing the influence of charismatic opinion-leaders,
6. who serve as visible role models of adoption for the non-adopters.

In this chapter, we broach an alternative conceptualization of diffusing innovations, which turns the classical diffusion paradigm on its head. This alternative approach to diffusing innovations is known as the Positive Deviance (PD) approach. The PD approach is not touted here as a substitute for the classical diffusion of innovations paradigm. Rather, we argue that the PD approach expands the solution space by working with a different set of principles, questions, and mindsets, believing that often the wisdom to solve intractable social problems lies within the community. Diffusion in the PD approach is an inside-out process in contrast with the classical dominant framework of outside-in diffusion.

The PD approach to diffusing "new ideas and practices" has been employed over the past two decades in over 40 countries to address a wide variety of intractable and complex social problems, including solving endemic malnutrition in Vietnam (Zeitlin, Ghassemi & Mansour, 1990; Sternin, J., 2003), decreasing neo-natal and maternal mortality in Pakistan (Shafique, Sternin & Singhal, 2010), reducing school dropouts in Argentina (Dura & Singhal, 2009), reintegrating returned child soldiers in northern Uganda (Singhal & Dura, 2009), drastically cutting down the spread of hospital-acquired infections in U.S. health-care institutions (Singhal, Buscell & McCandless, 2009), and in addressing many other issues (Pascale & Sternin, 2005; Pascale, Sternin & Sternin, 2010).

In this chapter we describe the Positive Deviance approach, including its key tenets and principles, by analyzing its historical origins in Vietnam to combat endemic malnutrition. Through the experience of this pioneering real-life application of PD in Vietnam, and drawing upon dozens of others that have followed, we argue for an alternative conceptualization of diffusion of innovations—one that turns upside down our cherished conceptualizations of expert and outside change agents, the notion of filling KAP gaps, the traditional role of opinion leaders, and the like.

What Is Positive Deviance?

Positive Deviance (PD) is *an approach to social change that enables communities to discover the wisdom they already have and then to act on it* (Sternin & Choo, 2000; Pascale & Sternin, 2005; Singhal & Dura, 2009). PD initially gained recognition in the work of Tufts University nutrition professor Marian Zeitlin in the 1980s, when she began focusing on why some children in poor communities were better nourished than others (Zeitlin, Ghassemi & Mansour, 1990). Zeitlin's work privileged an assets-based approach, identifying what's going right in a community in order to amplify it, as opposed to focusing on what's going wrong in a community and fixing it.

Jerry Sternin, a visiting scholar at Tufts University, and his wife, Monique Sternin, built on Zeitlin's ideas to organize various PD-centered social change interventions around the world. They institutionalized PD as an inside-out diffusion of innovations approach by showing how it could be operationalized in a community setting (Papa, Singhal & Papa, 2006).

Combating Malnutrition in Vietnam[1]

Location: Hanoi, Vietnam. December 1990

"Sternin, you have six months to show results," noted Mr. Nuu, a high-ranking official in the Vietnamese Ministry of Foreign Affairs.

"What? Six months? Six months to demonstrate impact?" Jerry Sternin could not believe his ears.

"Yes, Sternin, six months to show impact, or else I will not be able to extend your visa."

In December 1990, Jerry Sternin, accompanied by his wife, Monique, and 10-year old son Sam, arrived in Hanoi to open an office for Save the Children, a U.S.-based NGO. His mission: to implement a large-scale program to combat childhood malnutrition in a country where two-thirds of all children under the age of five suffered from malnutrition.

The Vietnamese government had learned from experience that results achieved by traditional supplemental feeding programs were not sustainable. When the programs ended, the gains usually tapered off. The Sternins had to come up with an approach that enabled the community, without much outside help, to take control of its nutritional status.

And quickly! Mr. Nuu had given the Sternins six months!

Crisis or Opportunity

From years of studying Mandarin, Jerry knew that the Chinese characters for "crisis" are represented by two ideograms: danger and opportunity. Perhaps there was an opportunity to try something new in Vietnam.

機 = 危 + 机

Crisis = Danger + Opportunity

Necessity is the mother of invention. If old methods of combating malnutrition would not yield quick and sustainable results, the Sternins wondered if the construct of Positive Deviance, coined a few years previously by Tufts University nutrition professor Marian Zeitlin, might hold promise.

Zeitlin broached the notion of positive deviance as she tried to understand why some children in poor households, without access to any special resources, were better nourished than others. What did they know, and what were they doing that others were not? Perhaps combating malnutrition called for an assets-based approach, that is, identifying what's going right in a community and finding ways to amplify it, as opposed to the more traditional deficit-based approach of focusing on what's going wrong in a community and fixing it.

Positive Deviance (PD) is an approach to social change that enables communities to discover the wisdom they already have, and then to act on it.

Positive deviance sounded good in theory. But no one, to date, had operationalized the construct to actually design a field-based nutrition intervention. Might it work in a community setting? How? The Sternins had no roadmaps or blueprints to consult. Where to begin?

Childhood malnutrition rates were high in Quong Xuong District in Than Hoa Province, south of Hanoi, where the Sternins had set up base. The Ho Chi Minh trail, the major supply route for the Vietcong guerillas during U.S. hostilities in Vietnam, snaked through Quong Xuong, and so suspicion of Americans was palpably high. The Sternins first task was to build trust with community members. The rest would follow.

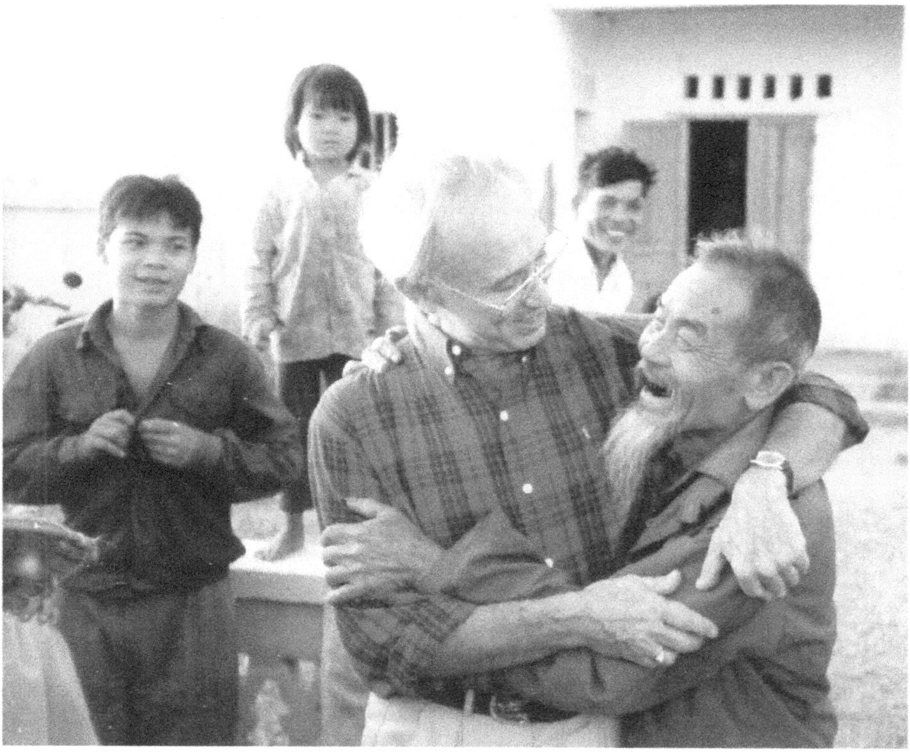

Building Trust: Jerry Sternin with village elder in Quong Xuong District, Vietnam.

After several days of consultation with local officials, four village communities were selected for a nutrition baseline survey. Armed with six weighing scales and bicycles, health volunteers weighed some 2,000 children under the age of three in four villages in a record 3.5 days. A growth card for each child, with a plot of their age and weight, was compiled. Some 64% of the weighed children were found to be malnourished.

No sooner were the data tallied, with bated breath the Sternins asked:

> "Are there any well-nourished children who come from very, very poor families?"

The response:

> "Yes, yes, there are some children from very, very poor families who are healthy!"

"Yes, yes, there are some children from very, very poor families who are healthy!"

These poor families in Than Hoa that had managed to avoid malnutrition without access to any special resources; these families would represent the Positive Deviants. "Positive" because they were doing things right, and "Deviants" because they engaged in behaviors that most others did not.

What behaviors were these PD families engaging in that others were not? To answer this question, community members were tasked to visit with six of the poorest families with well-nourished children in each of the four villages. The Sternins believed that if the community self-discovered the solution, they were more likely to implement it.

If the community self-discovered the solution, they were more likely to implement it.

Palpable excitement bathed the community hall. The self-discovery process yielded the following key PD practices[2] among poor households with well-nourished children:

- Family members collected tiny shrimps and crabs from paddy fields, adding them to their children's meals. These foods are rich in protein and minerals.
- Family members added greens of sweet potato plants to their children's meals. These greens are rich in beta carotene, and other essential micronutrients, e.g., iron and calcium.

- Interestingly, these foods were accessible to everyone, but most community members believed the foods were inappropriate for young children. Further,
- PD mothers were feeding their children three to four times a day, rather than the customary twice a day.
- PD mothers were actively feeding their children, making sure there was no food wasted.
- PD mothers washed the hands of the children before and after they ate.

> *When Positive Deviant practices are made visible, they are immediately actionable because they are accessible to everyone in the community.*

Doing Not Telling

With the "truth" discovered, the natural disposition urge was to go out and tell the people what to do. Now the "best practices" needed to be diffused so that the non-adopters could adopt them.

Various ideas for "telling" were brainstormed: household visits, attractive posters, educational sessions, and others. Many were implemented in the classical diffusion of innovations approach, trying to persuade people to see the relative advantages of these identified best practices. However, results were disappointing. While a few folks adopted the said best practices, the majority did not.

From their previous field-based experience in other countries, the Sternins knew that old habits die hard; new ones, even when they hold obvious advantages, are difficult to cultivate. The Sternins' experience suggested that such "best practice" innovations almost always engendered resistance from the people. The Sternins coined a phrase for it: the "natural human immune" response.

As the brainstorming winded down, a skeptical village elder bellowed: "A thousand hearings isn't worth one seeing, and a thousand seeing isn't worth one doing."

"A thousand hearings isn't worth one seeing, and a thousand seeing isn't worth one doing."

On the car ride back to Hanoi, the Sternins talked about the wisdom inherent in the elder's remark. Could they help design a nutrition program that emphasized "doing" more than "seeing" or "hearing?"

A two-week nutrition program was designed in each of the four intervention villages. Mothers, whose children were malnourished, were asked to forage for shrimps, crabs, and sweet potato greens. Armed with small nets and containers, mothers waded the paddy fields picking up tiny shrimps and crabs. The focus was on action, picking up the shrimps and crabs, and shoots from sweet potato fields.

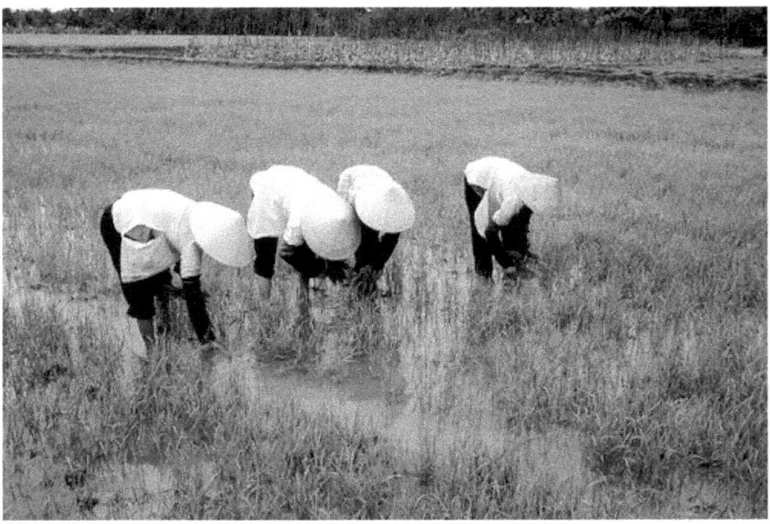

Shrimps and crabs for the taking in Vietnamese rice paddies.

"The PD approach is premised on the notion that it is easier to act your way into a new way of thinking than to think your way into a new way of acting."

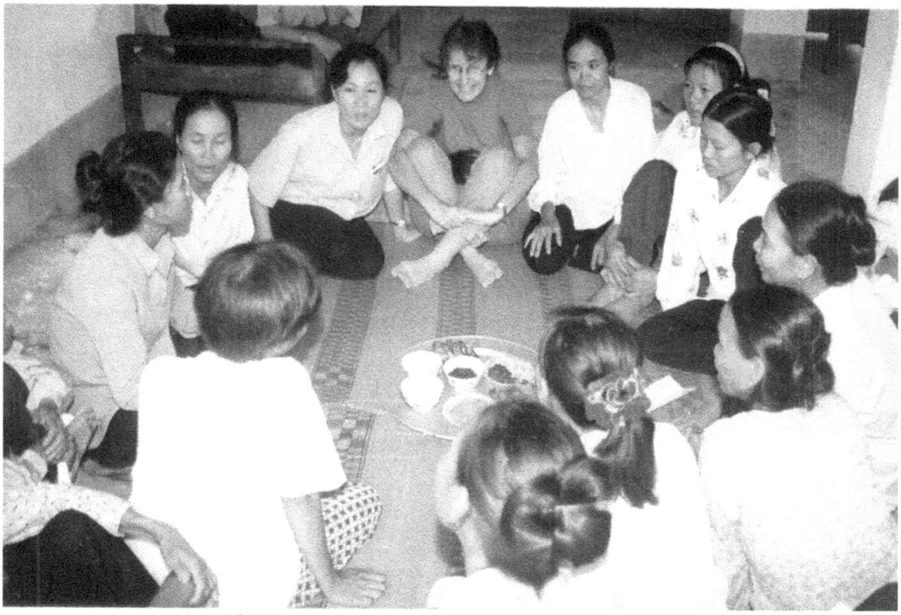

A cooking session in progress in an intervention village.

In the company of positive deviants, mothers learned how to cook new recipes using the foraged ingredients. Again, the emphasis was on "doing," on practice. Before the mothers sat down to feed their children, they weighed their children and plotted the data points on a growth chart. The children's hands were washed, and the mothers actively fed the children, ensuring no food was wasted. Some mothers noted how their children seemed to eat more in the company of other children. When returning home, mothers were encouraged to break the traditional two-meal-a-day practice into three or four portions.

Such feeding and monitoring continued for two weeks. Mothers could visibly see their children becoming healthier. The scales were tipping. And the rest is history.

The Positive Deviance approach is completely informed by, and bathed in, data. Data are collected at every step of the way and openly posted for the community members to monitor progress. Data informs where problems and the solutions lie.

After the pilot project, which lasted two years, malnutrition had decreased by an amazing 85% in the communities where the PD approach was implemented. Over the next several years, the PD intervention became a nationwide program in Vietnam, helping over 2.2 million people, including over 500,000 children, improve their nutritional status (Pascale, Sternin & Sternin, 2010; Pascale & Sternin, 2005; Singhal, Sternin & Dura, 2009; Singhal & Dura, 2009).

Born out of necessity, this pioneering PD experience in Vietnam turned the fundamental tenets of the classical diffusion of innovations framework on its head (Table 1).

Table 1: Positive Deviance approach contrasted with the diffusion of innovations approach.

Diffusion of Innovations Approach	Positive Deviance Approach
Solutions reside outside	Solutions exist within the community
Change agentry as pushing solutions	Community self-discovers solutions
Seeking adopter buy-in	Seeking community ownership
Emphasizing innovation attributes (relative advantage, compatibility, non-complexity, trialability, and observability)	The solution, by definition, delivers better outcomes (relatively advantageous), is compatible, non-complex (as people with no special resources have adopted). Further, the PD behaviors are trialable (already being practiced), and their results are observable. Now.
Expert change agents give advice	Change agents relinquish expertise, listen, and facilitate
Focused on plugging deficits	Focused on identifying and amplifying assets
Moves from problem-solving to solution identification	Moves from solution-identification to problem-solving
Adopters are persuaded	Adopters learn by doing
Susceptible to adopter resistance on account of exogenous solution	Open to self-replication on account of endogenous wisdom
Valorizes charismatic opinion leadership	Valorizes behaviors of ordinary people
Involves lengthy diffusion planning	Can begin now as solution resides in the now
Needs a heavy investment of resources for dissemination	Needs limited resources as someone is practicing those behaviors against all odds

Source: Draws upon Pascale & Sternin (2005), Singhal and Dura (2009), and Singhal, Sternin & Dura, 2009.

Since the Vietnam initiative, in the past two decades, the PD approach has been applied in a variety of contexts, to address a variety of intractable social problems, with highly effective outcomes (Pascale, Sternin & Sternin, 2010; Singhal & Dura, 2009). A growing body of literature validates the alternative perspective of inside-out diffusion as noted by the attributes of the PD approach in the above table.

Conclusions

The classical diffusion paradigm has been criticized for reifying expert-driven, top-down approaches to address problems and thus, by default, overlooking and rejecting local solutions (Papa, Singhal & Papa, 2006; Singhal & Dearing, 2006; Singhal & Dura, 2009). Diffusion of innovation experts now increasingly believe, and humbly acknowledge, the value of local expertise and indigenous wisdom in finding culturally appropriate solutions to community problems. One such inside-out approach to innovation diffusion is exemplified by the positive deviance approach.

The PD approach believes that innovations (or solutions) that are generated locally are more likely to be owned by the potential adopters. When adopters are externally persuaded to buy into the vision of an outside expert, they tend to demonstrate inertia and resistance, much like the Iowa farmers in the Ryan and Gross (1943) study who for an average of about 10 years resisted the adoption of hybrid seed corn.

The PD approach questions the traditional role of outside expertise, believing that the wisdom to solve the problem lies inside. While social change experts usually make a living discerning community deficits, and then implementing outside solutions to change them, in the PD approach, the role of experts is framed differently. The expert's role is to help the community find the positive deviants, identify their uncommon but effective practices, and then to design a community intervention to make them visible and actionable.

In the PD approach, the change is led by internal change agents who, with access to no special resources, present the social behavioral proof to their peers. If they can do it, others can, too. As the PD behaviors are already in practice, the solutions can be implemented without delay or access to outside resources. Further, the benefits can be sustained, since the solution resides locally.

Perhaps, most important, the PD approach turns the dominant "transmission-centered" innovation-decision framework on its head. As opposed to subscribing to the notion that increased knowledge changes attitudes and attitudinal changes change practice, PD believes in changing practice. PD believes that people change when that change is distilled from concrete action steps.

Notes

1. The author thanks the Positive Deviance Initiative at Tufts University, and particularly PD and diffusion practitioners and scholars with whom I have had the privilege of dialoguing over the past several years, including Monique Sternin, the late Jerry Sternin, and the late Everett M. Rogers, colleagues interested in the science of complexity such as Curt Lindberg, Henri Lipmanowicz, Prucia Buscell, and Keith McCandless, and UTEP colleagues Lucia Dura, Bobby Gutierrez, and others.
2. This pioneering Vietnam story draws upon numerous conversations and audiotaped interviews with Monique and the late Jerry Sternin, and partially from a co-authored case study (Singhal, Sternin & Dura, 2009).
3. A positive deviance inquiry focuses on eliminating those client behaviors from the strategy mix that are true but useless (TBU). TBU is a sieve through which a facilitator passes the uncommon qualities of positive deviants to ensure that the identified practices can be practiced by everyone.

References

Coleman, J. S., Katz, E. & Menzel, H. (1966). *Medical innovation: Diffusion of a medical drug among doctors*. Indianapolis, IN: Bobbs-Merrill.

Dearing, J. W., & Meyer, G. (2006). Revisiting diffusion theory. In A. Singhal & J. W. Dearing, *Communication of innovations: A journey with Ev Rogers* (pp. 29–60). Thousand Oaks, CA: Sage.

Dura, L., & Singhal, A. (2009). Will Ramon finish sixth grade? Positive Deviance for student retention in rural Argentina. *Positive Deviance wisdom series, number 2*, pp. 1–8. Boston, MA: Tufts University, Positive Deviance Initiative.

Frost, R. (1942). The secret sits. A poem in *The Witness Tree*. http://wonderingminstrels.blogspot.com/2001/01/secret-sits-robert-frost.html. Retrieved June 8, 2010.

Papa, M. J., Singhal, A., & Papa, W. H. (2006). *Organizing for social change: A dialectic journey of theory and praxis*. Thousand Oaks, CA: Sage.

Pascale, R. T., & Sternin, J. (2005). Your company's secret change agents. *Harvard Business Review*. May, 1–11.

Pascale, R. T., Sternin, J., & Sternin, M. (2010). *The power of positive deviance: How unlikely innovators solve the world's toughest problems*. Boston, MA: Harvard University Press.

Rogers, E. M. (2003). *Diffusion of innovations* (5th ed.). New York: Free Press.

Rogers, E. M. (2004). A prospective and retrospective look at diffusion model. *Journal of Health Communication*, 9(1), 13–19.

Rogers, E. M., & Kincaid, D. L. (1981). *Communication networks: A new paradigm for research*. New York: Free Press.

Ryan, B., & Gross, N. C. (1943). The diffusion of hybrid seed corn in two Iowa communities. *Rural Sociology*, 8, 15–24.

Shafique, M., Sternin, M., & Singhal, A. (2010). Will Rahima's firstborn survive overwhelming odds? Positive Deviance for maternal and newborn care in Pakistan. *Positive Deviance wisdom series, number* 5, pp. 1–12. Boston, MA: Tufts University, Positive Deviance Initiative.

Singhal, A., & Dearing, J. W. (Eds.). (2006). *Communication of innovations: A journey with Ev Rogers*. Thousand Oaks, CA: Sage.

Singhal, A., Buscell, P., & McCandless, K. (2009). Saving lives by changing relationships: Positive Deviance for MRSA prevention and control in a U.S. hospital. *Positive Deviance wisdom series, number 3*, pp. 1–8. Boston, MA: Tufts University, Positive Deviance Initiative.

Singhal, A., & Dura, L. (2009). *Protecting children from exploitation and trafficking: Using the Positive Deviance approach in Uganda and Indonesia*. Washington D.C.: Save the Children.

Singhal, A., Sternin, J., & Dura, L. (2009). Combating malnutrition in the land of a thousand rice fields: Positive Deviance grows roots in Vietnam. *Positive Deviance wisdom series, number 1*, pp. 1–8. Boston, MA: Tufts University, Positive Deviance Initiative.

Sternin, J. (2003). Practice positive deviance for extraordinary social and organizational change. In D. Ulrich, M. Goldsmith, L. Carter, J. Bolt, and N. Smallwood (eds.), *The change champion's fieldguide* (pp. 20–37). New York: Best Practice.

Sternin, J., & Choo, R. (2000). The power of positive deviancy. *Harvard Business Review*. January–February, 2–3.

Sternin, M., Sternin, J., & Marsh, D. (1999). Scaling up poverty alleviation and nutrition program in Vietnam. In T. Marchione (ed.), *Scaling up, scaling down* (pp. 97–117). New York: Gordon and Breach.

Valente, T. W. (1995). *Network models of the diffusion of innovations*. Creskill, NJ: Hampton Press.

Zeitlin, M., Ghassemi, H., & Mansour, M. (1990). *Positive deviance in child nutrition*. New York: UN University Press.

·10·

The Active Adopter in the Diffusion of Innovations

JAMES W. DEARING & GARY MEYER

Introduction

Within the diffusion of innovation paradigm, a user-centric understanding of adopters and implementers appropriately focuses on personal agency or activity. We theorize about adopter activity from the standpoint of the adopter, both in relation to other actual and potential adopters and in relation to the innovation in question. We use new formulations of diffusion theory from several different applied fields to make this case. We find that formulations of adopter activity are mostly relegated to social interaction and not interaction with an innovation. That is, one's innovation activity is easily and mistakenly conflated with an adopter's activity with other adopters such that one's level of activity is assumed to be defined as where one falls in the normal distribution for the trait innovativeness. Whereas the traditional approach to explaining diffusion positions adopters in a reactive role as socially connected receivers and evaluators of new ideas and things, new theories about diffusion suggest that adopters or their innovation implementers are often creators, too, in fully reworking innovations to achieve ideal compatibility between innovation and user context.

Many social theorists—especially those who concern themselves with mass communication effects—presume less-than-active roles for people. This tendency of behavioral theorists is not limited to scholars working in the 1930s, 1940s, and 1950s—that era when presumptions of strong and direct media effects constituted a dominant paradigm. Contemporary theorists, too, are attracted to models such as the social ecological framework that depict the individual as subject to multiple layers of imposing influences nested within one another. How, we may reasonably ask, can the isolated individual counteract let alone hold at bay all that occurs around and above him or her in forming and reinforcing attitudes, behaviors, and values? What chance has a person against those seemingly insurmountable forces of far-reaching communication technology, social networks, organizations, built environments, established norms, policies and regulations, social and class structures, nation-states, and the ubiquitous World Wide Web? When the techniques of mass marketing are so sophisticated in target audience segmentation and pinpoint tailoring, how does an individual ignore the message? What are the defenses against all that propels an individual to change?

The present chapter makes the case for a user-centric understanding of personal agency or *activity*. We do so with reference to diffusion of innovation theory, which continues to offer communication scientists a role to play in other, topically defined fields (Dearing, 2009). We begin this task by providing an example of how most diffusion theorists consider the process of diffusion to work in the case of programs and interventions designed for adoption by service intermediaries such as health-care professionals, community outreach workers, environmental managers, and schoolteachers (we shall refer to this generally accepted explanation as "the traditional approach"). After identifying the main components of the traditional approach, we then point to alternative ideas about diffusion on several fronts, led by scholars of management, public administration, and sociology. We argue that these divergent attempts to recast the diffusion paradigm share a common emphasis in conceptualizing what it is that potential adopters and actual adopters do in relation to one another and in relation to innovations.

Whereas the traditional approach to explaining diffusion positions adopters in a reactive role as socially connected receivers and evaluators of new ideas and things, new theories about diffusion suggest that adopters do more than alternate as conversationally active and reactive, depending on the innovation in question, in influencing others and being influenced by them. For despite the omnipresent influences to which individuals are subject, most potential adopters—that is, most of us—manage to reject innovations both passively and actively much more often than they respond positively. Potential adopters have defenses that they routinely employ against innovations. Furthermore, in those cases of acceptance, potential adopters can be full partners in the creative process of innovating as a result of actively framing the meaning of innovations, trading innovative ideas back and forth as innovation sources and innovation receivers, and by exploiting opportunities and constraints in the information and social environment.

We elucidate the broad multidisciplinary issue of adopter activity. Current formulations of adopter activity are mostly relegated to social interaction and not interaction with an innovation. That is, one's innovation activity is easily and mistakenly conflated with an adopter's activity with other adopters such that one's level of activity is assumed to be defined as where one falls in the normal distribution for the trait innovativeness. Our focus, instead, is on adopters as creators of innovations, a view of democratizing diffusion for public benefit that is strongly analogous to Eric Von Hippel's (2005) observation that innovating in multiple sectors is becoming a democratic process that, if recognized and embraced, allows for the capturing of private benefit. Stated differently, while the traditional view of the diffusion of innovations decouples the process of innovation from the process of diffusion, the theoretical perspective we propose here closely weds the two together for the purpose of detailing a means of "making a difference" in social conditions.

The particular context in which we assess adopter activity is the problematic and important multidisciplinary domain of translational research. *Translational research* is the study of how evidence-based practices, programs, and campaigns can best be communicated for adaptation by program staff and intermediaries for the benefit of their constituents. Translational research has become a critical concern of major research institutions and change agencies such as the U.S. National Cancer Institute. While the traditional approach to diffusion scholarship popularized by Katz (Katz & Lazarsfeld, 1956) and Rogers (1962) and since tested and explicated by them and others has applicable concepts and clear implications for advancing translational research, it is deficient in certain respects, too. Thus, we advocate a hybrid orientation to the intervention-based study of the diffusion of innovations, one that acknowledges (1) the enduring role that centralized change agencies will continue to play in seeking social betterment; (2) variance among program staff not just as adopters but as innovators, which we conclude is a type of decentralized diffusion; and (3) innovation specificity as a critical concern of program developers and their supporting change agencies in the pursuit of external validity and diffusion. We believe that our approach to the diffusion of innovations as translational research represents a unique niche, especially for scholars who consider the field of communication their home.

Diffusion As We Know It

How do we understand diffusion to work? Consider the following scenario:

> Min Wha has been an assistant professor of chemistry at a California State University campus for two years. She wants to demonstrate her willingness for professional development because her three-year review is only one year away, so Min Wha volunteers to attend a half-day workshop entitled Process Oriented Guided Inquiry Learning (POGIL), an approach to active learning developed by chemistry faculty at Franklin & Marshall College and the State University of New York at Stony Brook and funded by the U.S. National Science Foundation, hosted by her university's center for teaching excellence. The workshop is interesting to Min Wha. She finds POGIL easy to understand and learns that she can try it out little by little. Yet she decides nothing. Three weeks later she has a chance conversation in the mail room with Rukhsana, a more senior assistant professor whom Min Wha likes. Though she has never seen Rukhsana teach, Min Wha admires her colleague's teaching ability based on students' casual comments about what a good teacher Rukhsana is and the fact that Rukhsana won a university teaching award. In the mail room, Rukhsana talks as she recycles a flyer about constructivism as a basis for better teaching. Min Wha, alert, listens. She tells Rukhsana about the POGIL workshop. Rukhsana reveals that she, too, had attended a POGIL workshop and now organizes many of her classes around its approach to classroom learning. Back in her office, Min Wha decides to try POGIL with her chemistry students.

The preceding scenario demonstrates the accepted, traditional approach to explaining the individual decision process as described and explained via the concepts popularized in the book *Diffusion of Innovations* through its five editions over 44 years. Knowledge is gained largely through the one-way communication of information. Persuasion occurs through the two-way communication of normally operating social influence in the form of informal, extant local opinion leadership. Together, communication of information and communication of influence represents a dual-process model of inputs (Bandura, 1986) that can result in a ripple of positive adoption decisions among units of adoption such as teachers to try a new practice. In the preceding scenario, information alone was insufficient to move the individual toward a positive decision or even serious contemplation of innovation costs and benefits when the innovation in question is consequential (perceived to have high relevance to the attainment of personal or collective objectives). Talking was key. Proximity played a role. And it was not just anyone—a third faculty colleague, for example—who was able to prime Min Wha's memory in a productive way. It was Rukhsana, a person whom Min Wha already believed to be expert and trustworthy. If someone other than Rukhsana had advocated or served as a social model for problem-based learning, Min Wha likely would not make the decision she does.

Opinion leaders constitute only a small proportion of system members (often about 5%). So approaches to spread innovations by seeding them with randomly selected individuals will miss the mark in jump-starting diffusion about 19 times for every 20 attempts. More common approaches such as advertising for volunteers (self-selection), or assuming that people in positions of authority are also influential among near-peers may result in success rates better, or worse, than random selection of people. All three of these approaches of seeding an innovation in a social system are likely to be less effective at diffusing the innovation than one of several means of identifying sociometrically influential persons and recruiting them to model or talk about the innovation with potential adopters.

What else does the preceding scenario tell us about the traditional interpretation of how diffusion works? The adopter, Min Wha, is one in a network of like professionals. She does not only talk with Rukhsana. She is a teacher who talks with other teachers. Her relations with them are not evenly distributed, but there is a discernable structure to the network in which she is embedded and to which she contributes. Certain faculty she talks with more and regards more highly; others, less so. This patterned structure is rather topic-specific; in this instance, it is perhaps unique to teaching and learning expertise, though that structure may overlap, sometimes considerably, with related topics, such as chemistry research, university administration expertise, chemistry disciplinary knowledge, or industry consulting expertise. And by extension, we expect Min Wha to talk with students and chemistry professionals.

We can also note that Min Wha is initially prompted to action by an institutional condition: The looming requirement of promotion and tenure review. Institutional rules such as this one are a part of the environmental context within which individuals make decisions. Often, it is an environmental cue that precipitates individual action.

In the scenario, we see how change agencies and their agents of change enter into the information environments and social networks of potential adopters as innovation developers, the funders of their work, and a host intermediary (the center for teaching excellence). While adopters may be members of far-flung and loosely knit communication networks, the formal channels by which innovations are created, defined, tested, refined, and communicated are often centralized when obvious interested change agencies are involved. The traditional diffusion perspective is an over-time processual one, communication first occurring through more formal communication channels, from the few to the many, which gives way to interpersonal talk and the possibility of influence in informal channels.

This traditional understanding of diffusion also prominently features the importance of an adopter's impressions in leading to decisions. These attributes of innovations, codified by Rogers as complexity, relative advantage, compatibility, trialability, and observability, constitute perceived benefits and costs of a positive adoption decision. In our example, Min Wha develops a sense that it is easy to understand POGIL and that she need not make an all-or-nothing commitment to trying it with her classes of students.

Last, as the scenario depicts, the innovation in question, POGIL, is a specific entity with multiple components, not so much embedded in physical artifacts as it is a coordinated system of protocols, techniques, and approaches to accomplishing goal-directed objectives, like so many social programs and interventions. It was created by individuals who work at certain institutions, at a particular time. There are books, brochures, web sites, presentations, CD-Roms, and workshops all devoted to it. Potential adopters learn what POGIL is, the nature of the evidence supporting the workshop claims, how to implement POGIL, and how to get help with its use. Innovation developers are nearly always emotionally and sometimes financially vested in their work. They are the ultimate experts concerning the innovation in question. Because they have worked through many contingencies during the innovation development process, they resist its modification by others, justifiably uncertain that variants on what they have created will still be effective in achieving program objectives.

In summary, the traditional approach to understanding diffusion emphasizes (1) the key generative role of centralized change agencies and those they fund as the sources of innovation; (2) the importance of perceptions by potential adopters of the attributes of an innovation to decrease uncertainty about what the innovation is and can do; and (3) the necessity of personal influence for signaling potential adopters that an innovation should be attended to, that it is consonant with prevailing local social norms, and that it is promising. Our scenario is a rather delimited set of influences on an adoption decision; Wejnert (2002) provides an exhaustive listing of independent variables as either (1) characteristics of innovations; (2) characteristics of innovators; or (3) parts of the environmental context, each of which can affect adoption decisions.

Prevailing beliefs about how diffusion works are just that: related sets of concepts that have been the objects of study, in some cases, considerable study. Researchers working in anthropology, agriculture, sociology, education, and other disciplines have over seven decades rarely studied all aspects of the process of diffusion; rather, they have emphasized those concepts most germane to their discipline (Katz, Levin & Hamilton, 1963). Yet there are still other ways of explaining the spread of new ideas, even if they are less supported yet through empirical research.

Alternatives to Traditional Explanations of Diffusion

A Rhetorical Perspective on Diffusion

Green (2004) proposed a rhetorical explanation for the diffusion of managerial innovations. He argued that through their choices of language, managers increase or decrease the diffusion of new ideas and things within and across firms, such as Total Quality Management (TQM). The traditional approach to explaining diffusion includes the idea that when diffusion occurs, informational messages precede persuasive messages since potential adopters first have questions about what an innovation is (knowledge-based questions), and then later have questions about the innovation's performance and, in particular, its negative aspects (evaluation-based questions) as they move nearer to making an adoption decision. The traditional explanation also includes the idea that what potential adopters say to each other—and especially what opinion leaders say to near-peers—affects the rate of adoption of an innovation.

In a set of propositions, Green (2004) posited that diffusion rates will vary depending on whether justifications for adopting the innovation are based in pathos (emotion), logos (effectiveness), or ethos (morality). Pathos appeals, he argued, would result in rapid diffusion and rapid discontinuance; logos appeals in moderately rapid diffusion and moderately rapid discontinuance; and ethos appeals in slow uptake and gradual discontinuance. Green reasoned that by combining the three types of justifications in this sequence (pathos arguments, then logos arguments, followed by ethos arguments), diffusion would be both accelerated and prolonged. This formula is different from that in Rogers (2003), where persuasive appeals rich in pathos are understood to be necessary just prior to the time of adopter decision, and after what rhetoricians would consider logical or logos appeals to explain to potential adopters what the innovation is and how it works.

In Green's theorizing about organizations, it is the managers as potential adopters who control the determination of such meaning and its communication. The agency for change within the corporation is placed with the managers as adopters, and less so with the outward-to-inward efforts of change agencies such as consulting firms. While to Green, external change agencies remain the sources of innovation, the meaning of those tools, techniques, and technologies are fomented and negotiated internally, by adopters. To him as well as to others (Strang & Meyer, 1994; Abrahamson & Fairchild, 1999), discourse among adopters determines the fate of managerial innovations. This rhetorical perspective on diffusion is about framing the meaning of innovations. The theory re-conceptualizes how we think about innovations; it is a means of explaining how characterizations of innovations influence our adoption of them.

A Trading Perspective on Diffusion

On the basis of a series of empirical studies, several management scholars have developed the theory that individual adopters employed by competing organizations frequently engage in the informal and asynchronous exchange of proprietary knowledge to solve work-related problems. First postulated by Eric von Hippel (1987) and subsequently further modeled and tested by Carter (1989) and Schrader (1991), "know-how trading" has been shown to well characterize how engineers and other technical and professional staff learn of process innovations. Though these researchers do not term it as such, know-how trading is an enactment of social capital; adopters access valuable information through extant relational contacts on a need-to-know basis. Reciprocity is the guiding principle governing information exchanges through these relationships, just as it is with social capital of other types. What you help to solve this week for an acquaintance at a rival firm you will recoup at some unspecified later date. Schrader's (1991) study of 294 middle-level managers showed that the diffusion of knowledge through know-how trading benefited both the individuals who were party to the exchanges and their employers.

Traditional theorizing about diffusion strongly suggests a directionality of diffusion, from sources to adopters (receivers). Diffusion research that incorporates the study of social networks has demonstrated the mediating role of interpersonal relations in speeding or slowing rate of adoption (Coleman, Katz & Menzel, 1966). The theoretic importance of the trading perspective is its evidence of asynchronous reciprocity; that actors play the roles of sources and receivers at different times, according to the interpersonal request and supply for innovations. The implicit promise of engaging in exchange networks is the attendant obligation for reciprocal behavior, a strongly felt norm deriving from informal and tacit agreement. This sense of obligation, of brotherhood or sisterhood, facilitates bilateral diffusion, not just from A to B, but from B to C to A. The informality, multiplexity, and strength of interpersonal work-based relationships means that the high value information usually required for successful technology transfer—so-called tacit knowledge—is precisely the type of knowledge that know-how trading facilitates. The trading perspective on diffusion is about the maintaining of important relationships. Whereas the rhetorical explanation for diffusion is about the meanings given to ideas and things, this trading perspective on diffusion of practices and innovative solutions is about relationships, not what you think, but whom you know. Know-how trading re-conceptualizes how we think about adopters and the activities in which they engage. In know-how trading, adopters are mutually and maximally active. They do not merely reinvent innovations. They are the generative sources of innovative solutions to managerial and technical problems.

An Evolutionary Perspective on Diffusion

Douthwaite, Keatinge, and Park (2002) studied what agricultural adopters did with six post-harvest technologies in the Philippines and Vietnam. They concluded in a cross-case analysis that learning and cycles of adaptation—analogous, they argue, to natural selection in the biological world—characterized how adopters interacted with technologies as a result of their social and work environments. In particular, Douthwaite, Keatinge, and Park identified instances of novelty generation (the creation of differences between adopters, their innovations, or the ways that they use innovations) and the subsequent selection by adopters of the most beneficial "novelties" to best fashion the technology in question. The authors observed that repeated instances of novelty generation and selection of the most well-suited novelties led to results (innovations) with "hybrid vigor," innovations that were most able to thrive in use by the adopters in question. The fittest innovations survived from plausible promise to successful adaptation in practice.

In this approach to understanding technological change, innovation and diffusion are inextricably intertwined processes. But unlike the rhetorical perspective that it is justifications that drive diffusion or the trading perspective that it is relational obligation that drives the spread of innovation, the key force in the evolutionary perspective that determines the adaptations that adopters make to innovations is the environment, which both enables and constrains the prosperity of any one "species" or innovative solution. Adopters change innovations to exploit a perceived niche in which they want to operate. So this perspective re-conceptualizes how we think about environmental conditions in relation to diffusion, giving primacy to environmental conditions as a shaping force on innovation and propagation.

This theory suggests that while the environment determines what is possible, the agency necessary to take advantage of that opportunity is the adopter's, whose activity is as much defined by a process of innovating as it is by a process of diffusion. The evolutionary perspective on diffusion is about adjusting the innovation-immediate context fit so that an opportunity made possible by the environment can be exploited. The result, according to these authors, is highly tailored innovation, often unique to each individual adopter-user.

Adopter Activity as a Common Thread

While the managerial and agricultural technology explanations for diffusion just reviewed are quite distinct from one another, they have in common a claim that adopters of innovations are by no means passive receivers of innovations. Nor are adopters mere tinkerers or customizers, adapting at the margins of innovations from afar. These three disparate theories of diffusion suggest that adopters are not at all what we have been told they are. Rather, they are creators, inventors, and sources of change. They are, without overstating the case, extremely active in testing, manipulating, jerry-rigging, and doing what it takes to create, through language and through action, an innovation that precisely addresses the requirements of an acutely felt local problem. If the purpose at hand is to create something radical or wonderful, then the type of adopter activity to which we refer will be of small import. On the other hand, if the purpose is to heighten the utility of a given innovation and hence transform it from something that one will not adopt or use into something that suits a felt need well enough to cause both adoption and implementation, then adopter activity is vital. It is perhaps what Maslow (1959) referred to as *secondary creativity*, individual activity that accounts for the majority of creative if modest output "which are essentially the consolidation and exploitation of other people's ideas" (p. 93).

We find it helpful to distinguish two types of adopter activity. Potential adopters can be active in relation to other adopters and potential adopters. They can also be active in relation to an innovation. We discuss activity in relation to others first but then focus on the latter type of activity.

The traditional diffusion perspective is one in which potential adopters actively listen to, read about, and observe responses to innovations and discuss those innovations with others. People are neither passive nor atomized individuals as much communication scholarship has characterized mass audiences. Except for the most venturesome and the most cautious, potential adopters think and act with reference to the social norms that characterize the networks or systems of which they are members. Thus, for most people, their activity is of a social type that is normatively guided by prevailing attitudes and values. Their activity is with other adopters and potential adopters, defined more or less by each person's degree of innovativeness. The earliest adopters ("innovators" in Rogers' categorization) are highly active in scanning information environments, in seeking out new ideas from heterogeneous sources and in experimentation. Feeling few constraints on their behavior, they act nearly autonomously toward the group, though they often exhibit ties to others outside the immediate group. But innovators comprise only a small proportion (2.5%) of the adopters in any social system. The vast majorities of others (early adopters, early majority, late majority, laggards) are less active in how they behave, their activity being more a response to how they perceive that others within the group view the innovation in question. The last to adopt also exhibit a lesser degree of social integration, though they are more commonly passive rejecters rather than active in relation to others. The essential point is that activity is carried out through social relations. Conceptualizing one's activity as a property of interaction with others dominates network and structural perspectives on innovation diffusion (see, for example, Burt, 2003).

Adopter activity in relation to an innovation is much less prevalent in the diffusion literature. To be sure, there is support from studies that adopters actively change innovations during implementation (Rice & Rogers, 1980; Dearing et al., 1998). But activity of this sort has usually been considered deviant or minor in relation to the original source-defined purpose of an innovation and how it was designed by that source. Rogers (2003) and other scholars labeled such adopter activity *re-invention*. In the present chapter, we maintain along with Green (2004), Hippel (1987), and Douthwaite, Keatinge, and Park (2002) that adopter activity in relation to innovations is far more important, common, and consequential than reflected in the diffusion literature.

The Domain of Translational Research

For many social problems in the United States, the state of the science (what researchers collectively know about solutions to a given problem) and the state of the art (what practitioners collectively do to help their constituents with a given problem) co-exist more or less autonomously, each realm of activity often having little effect on the other (Dearing, 2003). This situation has been referred to as a "quality chasm" by the Institute of Medicine, a "problem of translation" by the National Cancer Institute and the Agency for Healthcare Research and Quality, and a challenge of "going to scale" by health policy entrepreneurs (Lenfant, 2003). Explicit calls for using diffusion theory to address problems of translation have been made (Berwick, 2003). By no means are health and health care the only sectors of U.S. society characterized by evidence-practice gaps, yet perhaps in no other sectors are discrepancies so well documented (Institute of Medicine, 2001).

Government agencies and philanthropic organizations have tended to see the closure of evidence-practice gaps as a problem of *doing* dissemination, that is, of distributing information about research results through familiar communication channels such as conferences, workshops, continuing education, proceedings, articles, and, increasingly, web sites. A different route to redressing evidence-practice gaps is the funding of research *about* dissemination, or what in certain federal agencies is termed translational research.

Translational research is the study of how evidence-based practices, programs, and campaigns can best be communicated to practitioners for the benefit of their constituents. When translational research strategies are based in diffusion of innovation concepts, the resulting study is a type of dissemination intervention that we label *purposive diffusion*.

A *practitioner* is someone who makes decisions about which programs and practices will be implemented to provide information, advice, support, and services to constituents. In organizations such as social service agencies, community-based nonprofits, schools, hospitals, and clinics, practitioners interact directly with constituents by counseling them or instructing them. Many such organizations receive contracts and grants from funders to carry out such programs and practices. For this reason, practitioner organizations are often referred to as intermediaries: They mediate and add value to the flow of resources from government and other funding sources to populations of people who need services. We seek to clarify the functional role of practitioners because the outcomes of translational research often focus on them. Their attitudes and behavior in response to the diffusion of innovations is key to eventual constituent outcomes (the percentage of low-income women of childbearing age who smoke goes down) and societal impacts (cancer incidence decreases among low-income women).

Translational research is a form of meta-intervention research in which the domain of action concerns how previously validated interventions can be spread through networks of practitioners who are interconnected through regular, patterned informal communication, or spread through sectors of practitioners who are not knowingly interconnected with each other but who are professionally, geographically, situationally, or behaviorally similar such that they attend to the same sources of information and thus can be reached via the same set of communication channels (the same web sites, the same trade association publications, the same conferences).

Translational research concerns the purposive spread of innovations that have been derived from empirical research and are repackaged for practitioners, for the benefit of the people served by those practitioners. Translation is challenging because the knowledge in question must traverse a heterogeneous divide, from the intellectual and social world of researchers on one side to the intellectual and social world of practitioners on the other. We know from many studies of failed technology transfer, failed knowledge dissemination, and failed research-to-practice that translation cannot be assumed. Researchers and practitioners often exhibit more heterogeneity in education, occupation, interests, and value systems. Practitioners tend to see themselves as members of client systems in which they attempt to gain entree and intervene on local cultural terms. They often live and work within the same communities that they try to affect. Researchers tend to be outside such systems, even when they spend long periods studying them, and are often identified more with change agencies, government departments, and universities. Effective practitioners are considered insiders by the people they try to help; researchers—even effective ones—rarely achieve such status. And while some types of information about effective practices can be communicated effectively and efficiently between heterogeneous groups, other types of information resist easy or rapid translation.

People retain some knowledge tacitly, not even knowing what it is that they themselves know and thus what it is that others, too, need to know. Tacit knowledge is sticky knowledge; it likes to stay where it is (Szulanski, 1996).

While these reasons for the common failure to share effective programs and practices between researchers and practitioners are understandable, conditions are actually worse. The perceptual basis of heterogeneity requires far less to manifest. Effective practices have been shown to require years to diffuse across departments *even within the same organization and even among similarly trained employees* (O'Dell & Grayson, 1998). While a sense of competition can lead people in the same positions but in different organizations to adopt innovations (Burt, 2003), it can also function within the firm as a disincentive on sharing and helping.

A focus on research-to-practice translation is new to many scholars of public health, public affairs, management, education, and other fields. For a behavioral scientist in forestry, for example, studying how the posting of markers along hiking trails may reduce degradation of forests would not be uncommon. But studying how proven practices in trail signage can be spread among forestry managers—a translational research topic—would be uncommon. To water conservation specialists, studies of novel approaches to water reclamation have been common. But studies of failed city to city adoption of evidence-based water reclamation projects—translation—are uncommon. And among disease prevention researchers, studies of hypodermic needle exchange programs for the reduction of HIV have proliferated. But translational studies of how to best encourage city health departments to adopt effective needle exchange programs have been rare.

Where are problems of translation vested? Who is responsible in the many cases when translation fails? Since research evidence in topical domains such as patient care and classroom learning is constantly produced and communicated, a lack of uptake by practitioners such as nurses and teachers is most often conceptualized as their problem, that is, a practitioner problem. The solution in typical dissemination terms is to communicate more, more often, in more ways. The highly publicized disparities in health care have been conceptualized and redressed in this way. If use of the most effective medical care options lag in the American south, medical staff there are blamed for not counseling patients to select those options (Greer et al., 2002). Translation is a problem of practitioners, and as a consequence it is their constituents who lose out.

The problems with this typical view of translation responsibility are twofold: First, solutions to this assignment of blame have not produced more uptake by practitioners. Doctors are blamed for not using a new effective procedure, so content about the procedure is added to continuing medical education modules *in addition to* e-mailed office alerts, messages to patient support groups for them to ask their doctors about the procedure, and professional conference presentations. Unfortunately, more communication, more often, in more ways does not predictably lead to more adoption and implementation of evidence-based practices. We know that practitioners are skilled at ignoring and disavowing the relevance of research to their problems of practice (Dearing et al., 1996). We know that evidence of effectiveness is not a strong predictor of adoption and use (Weiss, 2001). The diffusion literature is replete with ineffective innovations that are broadly diffused and of innovations that rapidly spread without any evidence of their effectiveness. The majority of studies of clinical practice guidelines, which recommend proven practices and procedures for patient conditions, document the difficulty of affecting the decisions of health care providers (Lomas, 1991) and should give pause to anyone with a more-information-is-better view of translation.

Second, researchers can be poor or unconcerned communicators (Dearing, Meyer & Kazmierczak, 1994); researchers can develop research-based interventions without consideration of practitioner needs or wants; and researchers can mistakenly perceive practitioners as homogeneous sets of professionals ("day care providers," "landlords") who do not have or experience personal, professional, workplace, and client-based variations that would logically lead to diverse practitioner preferences, too (Greer et al., 2002). In short, there is much to blame researchers for if part of the objective of the research in question is to make a difference among non-academics. In an attempt to sensitize researchers to some of these issues, an increasing number of funders of research have begun suggesting that "if we want more research-based practice, we need more practice-based research" (Orleans, 2004).

To summarize, adopters are conceived in much the same way in translational research as they are in diffusion studies. They are seen as somewhat active in relation to other adopter-practitioners and mostly passive in relation to communicated intervention programs. The chief difference is that in translational studies the adopters are themselves mediators in the process of change—they are practitioners—who work upstream of their clients, residents, members, and employees. They make adoption decisions about which programs they will implement to benefit their constituents. Unlike many of the constituents to whom public information campaigns and intervention programs are targeted, practitioners in many sectors are highly educated, subject to much outreach and professional information, technically sophisticated, and can be counted on to understand and use advanced communication technologies. For translation strategies, these differences have implications for how and what we choose to communicate with them.

The Active Adopter in the Process of Translation

What it is that adopters and implementers do with innovations has been viewed as a dichotomy. Either they adopt a practice or intervention as is, or they change it to better fit their current workplace or client conditions. Designers of interventions have come to believe that adaptation is either good or bad. For decades in discussions of how to best diffuse or "scale-up" effective educational programs, researchers have kept to this framing of the translational problem (Hutchinson & Huberman, 1993): Adherents of program fidelity believe that working to ensure that adopters make as few modifications as possible is key to retaining the success of the original program. If the program is changed, how does one know if it is still effective? On the other hand, adherents of the program adaptation perspective counter that it is only through allowing adopters to change a program to suit their needs that the likelihood of sustainability is increased. If adopters do not feel ownership of the program, how can we ensure its persistence in practice? Currently, the same debate is alive and well in disease-prevention circles (Backer, 1995; Elliott & Mihalic, 2004).

Our position in this debate is that designing intervention programs to resist modification is futile; the baby may be tossed with the bath water since adopters are free to look elsewhere for workable programs, or they will self-invent by borrowing from what their experience suggests will work and what impresses them from different sources. There is simply too much incentive at the individual or single organizational level to customize, to partly adopt, and to combine innovation components from elsewhere (Hippel, 2005).

First of all, the adaptation perspective of encouraging sources to enable adopters to change innovations is incomplete. More than an innovation must change if a best fit between a program and one's work context is to be achieved. The context, too, should change. If one only changes an adopted program and not the work environment—or vice versa—technical, delivery system, and performance criteria misalignments are likely to characterize implementation. Over-time and incremental adjustments to both an innovation and a work environment characterize successful cases of technology transfer (Leonard-Barton, 1988). "Mutual adaptation" of a new program and of its user environment implies that an awful lot of the action of successful diffusion occurs not with the source nor with the end-user such as a student, but in intermediary organizations such as a school. How practitioners interpret the purpose and promise of a new program will interact with how they choose to make accommodation for it in the workplace. The meanings we make of a new program will contribute to what changes in the workplace we deem useful to best exploit the innovation. Virtually all such user activities would be considered moderating factors in the process of diffusion; in translational research, however, these moderating variables assume center stage and should be given more importance in research designs as outcomes of translation (Glasgow, Lichtenstein & Marcus, 2003).

Second, adaptation is likely to intensify the meaning of an innovation for users since their actions in accommodation bring the innovation and its work environment more into harmony in achieving objectives that are important to each practitioner. Particularly if meaning is personalized and internalized, implementation could be greatly strengthened. Research has shown that people vest certain objects with sacred meaning. The attribution of sacred meaning to objects has been shown to occur across many types of people and across many types of objects. Marketing researchers Belk (1988) and Belk, Wallendorf, and Sherry (1989) argue on the basis of extensive data (Belk, 2005) that in attributing sacred meaning to objects, those objects become much more than possessions. They come to constitute an extended self of the individual. Key to this perspective is the idea that the meanings given are the product of the individual adopter, not of the company, change agency, or marketer selling or promoting the object. We speculate that the assignment of meaning to objects, especially sacred meaning, derives from adopter activity, such as coveting, trying out, learning about, playing with, and, especially, changing the object to most attach or create sacred meaning.

We suggest that even seemingly mundane innovations such as a sexually transmitted disease prevention program can become involving and personally meaningful to an adopting practitioner such as a community health outreach worker on the basis of the strong normative beliefs of the people in such positions of helping clients at risk of some dread occurrence. Feedback from representative samples of potential adopters to gather their perceptions of innovation complexity, relative advantage, compatibility, trialability, observability, and other relevant attributes prior to communicating can be used in the design of communication materials to heighten the likelihood of positive responses. But all such adjustments must be balanced with the need for effect fidelity. In the face of much adaptation, how can the likelihood of effective outcomes as the result of fielding a program be increased?

Strategies for Effective User Activity

We begin with the premise that internal and external validity are important concepts that should be understood and considered by innovation developers and implementers. *Internal validity* is the extent to which a causal inference may be made between two or more variables (Campbell & Stanley, 1966; Cook & Campbell, 1979). In other words, internal validity is focused on whether or not program components cause specific and desired program outcomes. In practice, consideration of internal validity is often framed as "Does the program work?" Internal validity is established through program testing under controlled conditions. The more highly controlled the environment within which a program is tested, the more confidence in the validity of the causal relationship(s) established.

External validity refers to the extent to which the presumed causal inference may be generalized to other populations, settings, treatment variables, or measurement variables (Campbell & Stanley, 1966). The objective of external validity is not simply *scale-up*, the replication of internal effectiveness in similar contexts. Lee J. Cronbach (1982), who led in the 1980s in theorizing about external validity theory, suggested that the real aim of external validity tests should be replication of effect across different contexts. The more an intervention that was demonstrated and validated in one site can still achieve positive effects when translated in other places, other populations, other times, and other topics, the more externally valid the intervention. So *external validity* is the degree to which a program shown to be effective in an initial test subsequently produces similar outcomes in dissimilar contexts. In Guala's (2003) words, " . . . it [the program] is *externally* valid if A causes B not only in E, but also in a set of other circumstances of interest F, G, H, etc." (p. 4, italics in the original). Consideration of external validity is often framed as "Does the program work *here*?"

We turn our attention to the application of internal and external validity in purposive diffusion given the premise that the most active users of innovations will be those who modify the innovation to best fit their circumstance and modify their circumstance to best fit the innovation (Leonard-Barton, 1988). Consider first the topic of internal validity. Rather than posing the question "Does the program work?" perhaps the question can be reframed as "Why does the program work?" or "What are the core components of the program that are thought to be integral to its effectiveness?" Rogers (2003) refers to this as principles knowledge or "information dealing with the functioning principles underlying how an innovation works" (p. 173).

A potential adopter who understands specifically why a program works will be better able to determine those aspects or core components of the program or environment that may be modified with minimal risk to program effectiveness. We suggest that user activity will produce innovations of higher external validity to the extent that core components are known and understood. Developers may tailor communication efforts to explain why or how the program works. Precisely how to do this should not be assumed or guessed at, but rather derive from careful formative evaluation to assess potential adopter needs and wants, and their reactions to successive prototypes of the innovation.

In addition to communicating the conceptual basis for why a program works, developers may communicate in several alternative ways to bring about a desired outcome. Consider, for example, a violence prevention program focused on building self-efficacy as a core component wherein increased self-efficacy is determined to be causally related to decreased violent behavior. In this case, developers may provide potential adopters a set of five different options for increasing self-efficacy (one innovation with five examples of how to replicate the desired effect). This gives the practitioner flexibility as well as guidance to adapt the intervention in a fashion most appropriate to the circumstance with little risk to fidelity. Such a strategy emphasizes effect fidelity (the achievement of the source's desired effects in subsequent external validity tests of the program) over program fidelity (exact replication of the program as it was originally demonstrated). Developers can enhance productive user activity by ensuring that users understand the underlying premise for desirable effects and, wherever possible, by providing users with alternative means to achieve the same effect.

Turning to external validity, the question "Does the program work *here*?" may be reframed as "How are aspects of the circumstance and environment here similar to and different from the circumstance and environment of the demonstration?" This question incorporates two principles identified by Cook (2000) and Shadish, Cook, and Campbell (2002) through which the external validity of specific programs can be assessed. The first relates to the extent to which prototypical characteristics of a model program are like those in second-order sites (what they term "surface similarity") and the second focuses on clarifying which differences between the original demonstration and the subsequent second-order tests are superfluous ("ruling out irrelevancies"). Stated differently, an important related means for communicating the core components that were causally responsible for the internal validity of the program is to also portray the peripheral components that were a part of the original demonstration of the program but determined to be not causally important, as well as the complementary assets of the organizations involved that implemented the program. Peripheral components of a program are those that adopters can be encouraged to change. And the listing of complementary assets or organizational capacities related to fielding the program can give subsequent adopters a strong sense of what it took to successfully implement the program in question.

We suggest that user activity will be relatively more effective if careful thought is given to important similarities as well as important differences between demonstration sites and adopting sites. Developers might, for example, communicate to intermediaries about the subjects used in the demonstration as well as other groups of subjects thought to benefit from the program. At the same time, other groups of subjects to whom the program should not be administered can also be pointed out. Similarities and dissimilarities might also be communicated about program delivery, timing, location, and so forth. With such guidance, intermediaries will be better able to make informed decisions about which aspects of the program or their environment should be changed to achieve a best fit between the two. Of course, some of the responsibility must fall to potential adopters as well to make reasonable determinations concerning the extent and importance of certain similarities and differences between their site and prior test sites.

Another promising strategy for increasing the likelihood that adopter-implementers will be able to come to a "best fit" of innovation and work environment is the grouping of several effective innovations together in an alternative innovation cluster. Having a delimited set of choices—a few logical alternatives—as a basis for decision making is cognitively appealing. If there are many choices, people often will not decide anything; not having a ready comparison similarly decreases the likelihood of making a selection (Schwartz, 2004). An *alternative innovation cluster* is a set of effective innovations that comprise different means to achieve the same or similar end. The clustering of alternatives is attractive for two reasons. First, change agencies often have catalogues of many interventions of a type, each created by grantees, each of which often addresses the same problem. Piecing together an alternative innovation cluster of effective innovations does not put the change agency in the position of "picking a winner" and run the risk of seemingly advocating one program at the expense of other effective solutions. An excellent example of this strategy of packaging alternatives is the U.S. Centers for Disease Control and Prevention's Diffusion of Effective Behavioral Interventions (DEBI) Project (Collins, Harshbarger, Sawyer & Hamdallah, 2006). Each intervention is a proven means of reducing the risk of HIV/AIDS in U.S. communities.

As with the DEBI Project, clustering enables potential adopters choice from among a set, thus giving them a way to become active participants in the weighing, matching, and then selecting of an intervention. We can also expect that in many cases of subsequent implementation, adaptation of the intervention and perhaps of the host organization will occur. Helping implementers to responsibly adapt so that they can realize optimum results will not only be positively related to adoption but also to sustained use by intermediaries. Their activity need not be feared; it can be positively related to adoption, implementation, and sustainability of change (Douthwaite, 2002). New technologies increasingly allow for vivid side-by-side portrayals of innovations and enable users to become more active as innovators in the pursuit of mutual and effective adaptation (Hippel, 2005).

Conclusion

An appreciation of the great and common extent of adopter activity with regard to the mutual adaptation of effective programs from afar and one's immediate work environment requires the acknowledgment that *every practitioner is unique.* We do not make this point to argue for the supremacy of idiographic orientations to knowing over the nomothetic; we fully appreciate the ability to find central tendencies among types of adopters. We emphasize uniqueness because it powerfully underscores the need to develop translational strategies that encourage choice, that acknowledge multiple routes to desired causality, and that provide scenarios and examples of the same principles but of different people and different contexts so that one can see and imagine the creative possibilities of application. We postulate that when practitioners exert creativity in implementation and those creative efforts are not just permitted but guided and encouraged by the initial developers of innovations and external change agencies, effective use of innovations will more commonly occur.

What we have argued here is a particular approach to decentralized diffusion. Like other approaches to decentralized diffusion such as those prevalent in the field of international development (Chitnis, 2005), it relies on high degrees of participation from the users of innovations. But there, perhaps, the similarity ends. We are strong believers in the importance and longevity of centralized change agencies in social betterment. Private foundations, international NGOs, and federal agencies will persist and grow. Our approach to decentralized diffusion is one that seeks to marry the advantages of centralized resources and the efficiency of diffusion effects, with good options for individuation at the level of practitioner-adopters. User needs exhibit high heterogeneity (Hippel, 2005); researchers and others who design the communication of effective programs to practitioners would be wise to heed this fact. We conclude that pursuing a purposive diffusion-based strategy to translation that has as its objective generalized causal inference (a fidelity of effect) rather than program fidelity (replication) is a most promising route to effective translation of research-to-practice.

References

Abrahamson, E., & Fairchild, G. (1999). Management fashion: Lifecycles, triggers and collective learning processes. *Administrative Science Quarterly*, *44*: 708–740.

Backer, T. (1995). Assessing and enhancing readiness for change: Implications for technology transfer. T. Backer, S. David & G. Soucy (eds.), *Reviewing the behavioral science knowledge base on technology transfer* (pp. 21–41). National Institute on Drug Abuse, Rockville, MD.

Bandura A. (1986). *Social foundations of thought and action: A social cognitive theory*. Englewood Cliffs, NJ: Prentice-Hall.

Belk, R. W. (1988). Possessions and the extended self. *Journal of Consumer Research*, *15*: 139–168.

Belk, R. W. (2005). Possessions, self, and the sacred. A. Griffin & C. C. Otnes (eds.), *16th Paul D. Converse symposium*. Chicago: American Marketing Association, 46–66.

Belk, R. W., Wallendorf, M., & Sherry, Jr., J. F. (1989). The sacred and the profane in consumer behavior: Theodicy on the Odyssey. *Journal of Consumer Research*, *16*: 1–38.

Berwick, D. M. (2003). Disseminating innovations in health care. *Journal of the American Medical Association*, 289: 1969–1975.

Burt, R. S. (2003). Social origins of good ideas. Unpublished manuscript. Chicago: Graduate School of Business, University of Chicago.

Campbell, D. T., & Stanley, J. C. (1966). *Experimental and quasi-experimental designs for research*. Chicago: Rand McNally.

Carter, A. P. (1989). Know-how trading as economic exchange. *Research Policy*, *18*: 155–163.

Chitnis, K. (2005). Communication for empowerment and participatory development: A social model of health in Jamkhed, India. Ph.D. Dissertation. College of Communication. Athens,

OH: Ohio University.

Coleman, J. S., Katz, E., & Menzel, H. (1966). *Medical innovation: A diffusion study*. New York: Bobbs-Merrill.

Collins, C., Harshbarger, C., Sawyer, R., & Hamdallah, M. (2006). The diffusion of effective behavioral interventions project: Development, implementation, and lessons learned. *AIDS Education and Prevention, 18*, Supplement A, 5–20.

Cook, T. D. (2000). Toward a practical theory of external validity. In L. Bickman (ed.), *Validity & social experimentation*. Thousand Oaks, CA: Sage, 3–43.

Cook, T.D., & Campbell, D.T. (1979). *Quasi-experimentation: Design & analysis issues for field settings*. Chicago: Rand McNally.

Cronbach, L. J. (1982). *Designing evaluations of educational and social programs*. San Francisco: Jossey-Bass.

Dearing, J. W. (2009). Applying diffusion of innovation theory to intervention development. *Research on social work practice*. OnlineFirst, June 5: 1–16.

Dearing, J. W. (2003). The state of the art and the state of the science of community organizing, T. Thompson, A. Dorsey, K. Miller & R. Parrott (eds.), *Handbook of health communication*. Mahwah, NJ: Lawrence Erlbaum, 207–220.

Dearing, J. W., Larson, R. S., Randall, L. M., & Pope, R. S. (1998). Local reinvention of the CDC HIV prevention community planning initiative. *Journal of Community Health*, 23: 113–126.

Dearing, J. W., Meyer, G., & Kazmierczak, J. (1994). Portraying the new: Communication between university innovators and potential users. *Science Communication*, *16*(1): 11–42.

Dearing, J. W., Rogers, E. M., Meyer, G., Casey, M. K., Rao, N., Campo, S., & Henderson, G. M. (1996). Social marketing and diffusion-based strategies for communicating health with unique populations: HIV prevention in San Francisco. *Journal of Health Communication*, *1*: 343–363.

Douthwaite, B. (2002). *Enabling innovation: A practical guide to understanding and fostering technological change*. London: Zed.

Douthwaite, B., Keatinge, J. D. H., & Park, J. R. (2002). Learning selection: An evolutionary model for understanding, implementing and evaluating participatory technology development. *Agricultural Systems*, *72*: 109–131.

Elliott, D. S., & Mihalic, S. (2004). Issues in disseminating and replicating effective prevention programs. *Prevention Science* 4(1): 47–53.

Glasgow, R. E., Lichtenstein, E., & Marcus, A. C. (2003). Why don't we see more translation of health promotion research to practice? Rethinking the efficacy-to-effectiveness transition. *American Journal of Public Health*, 93(8): 1261–1267.

Green, S. E. (2004). A rhetorical theory of diffusion. *Academy of Management Review*, *29*: 4: 653–669.

Greer, A. L., Goodwin, J. S., Freeman, J. L., & Wu, Z. H. (2002). Bringing the patient back in: Guidelines, practice variations, and the social context of medical practice. *International Journal of Technology Assessment in Health Care*, *18*(4): 747–761.

Guala, F. (2003). Experimental localism and external validity. Unpublished paper presented at the 2002 Philosophy of Science Association meeting.

Hippel, E. V. (1987). Cooperation between rivals: Informal know-how trading. *Research Policy*, *15*: 291–302.

Hippel, E. V. (2005). *Democratizing innovation*. Cambridge, MA: MIT Press.

Hutchinson, J., & Huberman, M. (1993). Knowledge dissemination and use in science and mathematics education: A literature review. NSF EHR/RED report 93–75. Arlington, VA: National Science Foundation.

Institute of Medicine. (2001). *Crossing the quality chasm: A new health system for the 21st century*. Committee on Quality of Health Care in America. Washington, DC: National Academy Press.

Katz, E., & Lazarsfeld, P. (1956). *Personal influence: The part played by people in the flow of mass communications*. Glencoe, IL: Free Press.

Katz, E., Levin, M. L., & Hamilton, H. (1963). Traditions of research on the diffusion of innovation. *American Sociological Review*, 28(2): 237–252.

Lenfant, C. (2003). Clinical research to clinical practice—lost in translation? *The New England Journal of Medicine*, 349(9): 868–874.

Leonard-Barton, D. (1988). Implementation as mutual adaptation of technology and organization. *Research Policy*, 17: 251–267.

Lomas, J. (1991). Words without action? The production, dissemination, and impact of consensus recommendations. *Annual Review of Public Health*, 12: 41–65.

Maslow, A. H. (1959). Creativity in self-actualizing people. H. H. Anderson (ed.), *Creativity and its cultivation* (pp. 83–95). New York: Harper.

O'Dell, C., & Grayson, C. J. (1998). If only we knew what we know: Identification and transfer of internal best practices. *California Management Review*, 40(3): 154–174.

Orleans, C. T. (2004). Oral presentation. Increasing physical activity in populations: Understanding diffusion and dissemination. The Cooper Institute, October 21–23. Dallas, TX.

Rice, R. E., & Rogers, E. M. (1980). Re-invention in the innovation process. *Knowledge*, 1: 499–514.

Rogers, E. M. (1962). *Diffusion of innovations*. New York: Free Press.

Rogers, E. M. (2003). *Diffusion of innovations*. New York: Free Press.

Schrader, S. (1991). Informal technology transfer between firms: Cooperation through information trading. *Research Policy*, 20: 153–170.

Schwartz, B. (2004). *The paradox of choice: Why more is less*. New York: HarperCollins.

Shadish, W. R., Cook, T., & Campbell, D. T. (2002). *Experimental and quasi-experimental designs for generalized causal inference*. Boston, MA: Houghton Mifflin, 341–373.

Strang, D., & Meyer, J. W. (1994). Institutional conditions for diffusion. In R. W. Scott & J. W. Meyer (eds.), *Institutional environments and organizations: Structural complexity and individualism*. Thousand Oaks, CA: Sage, 100–111.

Szulanski, G. (1996). Exploring the internal stickiness: Impediments to the transfer of best practice within the firm. *Strategic Management Journal*, 17: 27–43.

Weiss, C. H. (2001). What kind of evidence in evidence-based policy? Unpublished paper presented at the Third International, Inter-disciplinary Evidence-Based Policies and Indicator Systems Conference, University of Durham, UK.

Wejnert, B. (2002). Integrating models of diffusion of innovations: A conceptual framework. *Annual Review of Sociology*, 28: 297–326.

Contributors

GEORGE A. BARNETT is Professor of Communication at the University of California, Davis. His research interests focus on social and cognitive system dynamics and the examination of social and communication networks and how they change over time as a function of perturbations in the system in which they are embedded.

HARRY BOUWMAN is Associate Professor at Delft University of Technology and Privat Docent at Åbo Akademi University, Turku, Finland. His research interests are ICT and organizations, strategic management in relation to ICT management, business models, customer value and the development of 3G+ Mobile telecommunications services.

HAO CHEN is Assistant Professor of Communication Technologies at University of Wisconsin, Platteville. His research interests include multidimensional scaling system and neural network's application in the social science field, diffusion of innovation, analysis and information visualization of social networks, and measurement of attitude and behavior change.

JAMES DANOWSKI is Associate Professor of Communication at the University of Illinois at Chicago. He studies new media diffusion and the uses and effects of new media using his WORDij automated semantic network analysis software suite, social network analysis, and a variety of other methods.

JAMES DEARING is a behavioral researcher with Kaiser Permanente's Institute for Health Research in Colorado, where he directs the Cancer Communication Research Center, a National Cancer Institute-designated center of excellence in cancer communication research. He conducts studies of how innovations spread through advice networks of intermediary service providers, such as physicians and nurses.

BENJAMIN ELBIRT is an independent researcher with a Master's Degree from the Department of Communication at the University of Buffalo. His research interests include the study of network analysis, agent-based modeling and simulation, mathematical modeling, data visualization, animation and intonation and the application of these interests to real world social phenomena.

JULIA GLUESING is Research Professor at Wayne State University in the Industrial and Manufacturing Department, Detroit, Michigan. Her research interests include accelerating the diffusion of innovations in globally networked organizations, global virtual teams, collective action, mobile work, culture's influence on work, and ethnographic methods approaches to the study of networks and collaboration.

VERONIKA KARNOWSKI is a research assistant at the Institut für Kommunikationswissenschaft und Medienforschung at Ludwig-Maximilians-Universität in Munich. Her research interests lie in diffusion, appropriation, mobile communication, and new media.

KYOUNGHEE HAZEL KWON is a PhD candidate in communication at SUNY-Buffalo. She is interested in structural analysis of interpersonal, organizational, and computer-mediated communication. She is currently working on her dissertation, exploring social network effects on user behaviors in social network sites.

CAROLYN A. LIN, Professor of Communication, University of Connecticut, researches the content, uses, and effects of media, media technologies, advertising, health communication, and international communication. She's a Distinguished Research Faculty award recipient and the founder of the Communication Technology Division at the Association of Education in Journalism & Mass Communication.

GARY MEYER is Professor of Communication in the Diederich College of Communication at Marquette University. His research interests focus on diffusion theory and persuasion theory, both often applied within the areas of health promotion and disease prevention.

OSCAR PETERS is Managing Director at the Institute for Behavioral Research at the University of Twente, The Netherlands. His research interests are (mass) media use and effects, in particular the positive and negative effects of video-game play and the adoption and use of new media and ICT, with a special interest in the adoption of ambient technologies.

KEN RIOPELLE is Research Professor at Wayne State University in the Industrial and Manufacturing Department, Detroit, Michigan. His research interests include accelerating the diffusion of innovations in globally networked organizations, the study of collaborative innovation networks, or COINs, and the use of mixed methods research for analytical purposes.

ARVIND SINGHAL is the Samuel Shirley and Edna Holt Marston Endowed Professor of Communication, University of Texas, El Paso, and also appointed as the William J. Clinton Distinguished Fellow at the Clinton School of Public Service, Little Rock, Arkansas. He teaches and conducts research in the diffusion of innovations, organizing for social change, and the entertainment-education strategy.

FRANK TUTZAUER is Associate Professor of Communication at the University of Buffalo. His research interests include the study of conflict, particularly through game-theoretic approaches, network analysis, simulation, and mathematical modeling. Recent publications have appeared in *Communication Monographs*, *Social Networks*, and the *Journal of Mathematical Sociology*.

LIDWIEN VAN DE WIJNGAERT is Assistant Professor of Communication at the University of Twente. Her research interests focus on the adoption, use, and effects of ICT from a user perspective. Lidwien van de Wijngaert has studied the diffusion of a.o. Broadband Internet, HDTV, and Mobile Services. As of September 2008 she focuses on the diffusion of online public services.

THILO VON PAPE is Research Associate at the Chair for Interactive Media and Online Communication of the University of Hohenheim in Stuttgart, Germany. His research interests are diffusion and appropriation of new communication technologies, mobile communication, and persuasion through mobile and interactive media.

ARUN VISHWANATH is Associate Professor of Communication at the University of Buffalo. His research focuses on the cognitive, affective, and behavioral determinants of technology adoption. His research looks at how individuals conceptualize new innovations and the role of interpersonal and mass-mediated messages in the adoption and utilization of innovations.

WERNER WIRTH is Professor for Empirical Research at the Institute of Mass Communication and Media Research of the University of Zurich, Switzerland. His research interests are diffusion, adoption, and use of new media as well as cognitive, emotional, and persuasive aspects of media use, media reception, and media effects.

Index

www.ingramcontent.com/pod-product-compliance
Ingram Content Group UK Ltd.
Pitfield, Milton Keynes, MK11 3LW, UK
UKHW021827100926
8556IPUK00003B/203

9 781433 110832